GENETICS &
EVOLUTION

GENETIC
ENGINEERING

GENETICS &
EVOLUTION

GENETIC ENGINEERING

Manipulating the Mechanisms of Life

RUSS HODGE

FOREWORD **BY** **NADIA ROSENTHAL, PH.D.**

Facts On File
An imprint of Infobase Publishing

This book is dedicated to the memory of my grandparents, E. J. and Mabel Evens
and Irene Hodge, to my parents, Ed and Jo Hodge,
and especially to my wife, Gabi,
and my children—Jesper, Sharon, and Lisa—with love.

⚛

GENETIC ENGINEERING: Manipulating the Mechanisms of Life

Facts On File, Inc.
An imprint of Infobase Publishing
132 West 31st Street
New York NY 10001

Library of Congress Cataloging-in-Publication Data
Hodge, Russ, 1961–
   Genetic engineering : manipulating the mechanisms of life / Russ Hodge ;
foreword by Nadia Rosenthal.
      p. cm.
   Includes bibliographical references and index.
   ISBN-13: 978-0-8160-6681-0
   ISBN-10: 0-8160-6681-7
   1. Genetic engineering—Popular works. I. Title.
   QH442.H628 2009
   660.6'5—dc22   2008033700

Facts On File books are available at special discounts when purchased in bulk quantities
for businesses, associations, institutions, or sales promotions. Please call our
Special Sales Department in New York at (212) 967-8800 or (800) 322-8755.

You can find Facts On File on the World Wide Web at
http://www.factsonfile.com

Text design by Kerry Casey
Illustrations by Lucidity Information Design
Photo research by Elizabeth H. Oakes

Printed in the United States of America

Bang Hermitage 10 9 8 7 6 5 4 3 2 1

This book is printed on acid-free paper.

"I say that it touches a man that his blood is sea water and his tears are salt, that the seed of his loins is scarcely different from the same cells in a seaweed, and that of stuff like his bones coral is made. I say that the physical and biologic law lies down with him, and wakes when a child stirs in the womb, and that the sap in a tree, uprushing in the spring, and the smell of the loam, where the bacteria bestir themselves in darkness, and the path of the sun in the heaven, these are facts of first importance to his mental conclusions, and that a man who goes in no consciousness of them is a drifter and a dreamer, without a home or any contact with reality."

—from *An Almanac for Moderns:*
A Daybook of Nature
by Donald Culross Peattie
copyright © 1935 (renewed 1963)
by Donald Culross Peattie

Contents

Foreword

It is 2020, and a routine visit to the family doctor includes a report on your health generated with a battery of diagnoses based on your unique genomic fingerprint, produced with high throughput DNA sequencing from a small sample of your blood cells. Your doctor suggests remedial treatment for a defect in one of your genetic circuits that predisposes you to a lethal cancer that killed your grandfather, but in your case an RNA interference-based drug will reduce your susceptibility by knocking down the overexpression of the offending gene. Your doctor notes a characteristic single nucleotide polymorphism (SNP) in a gene encoding one of your metabolic enzymes that is associated with compromised kidney function later in life and suggests a diet rich in a fruit genetically engineered to provide you with the supplemental protein. You are not fazed by this arcane terminology; you understand the shorthand explanation of the medications your doctor prescribes, thanks to the standard education in modern molecular genetics you received in high school, now considered as important as a basic understanding of hygiene and nutrition.

Why study genetics today? It is clear that a revolution is under way that is rapidly transforming the fields of human health and whose consequences are already seen in the medical, pharmaceutical, and agricultural industries, changing our society in innumerable ways. The molecular pathways through which hereditary information flows are now common parlance in biology textbooks, enriching our understanding of human life and how it develops. With the human genome sequence in hand, scientists have the tools to map and characterize mutations along human chromosomes that cause hundreds of genetic illnesses and have generated model systems to study and design cures for many of them.

With the hope that these advances offer comes the hype. We are fascinated and perplexed by weekly announcements from the scientific community that flood the mass media, dazzling us with the prowess of stem-cell technology, for example, or raising concerns about the cloning of animals. Despite the importance of these new biotechnological advances, the general public is often misled by the aura of science fiction that often surrounds these discoveries and remains largely in the dark as to the actual consequences—the potential for good and the risk of harm. The impact of progress in genetic technologies is often misunderstood, even by some people involved in the research.

Hence the need for *Genetic Engineering,* a book that shows how genetics has become so important. This book chronicles the origins and fascinating history of genetic discovery and charts the current status and future promise of the field. Although a relatively young discipline with scarcely a century to its credit, genetics has gained a position of central prominence in the biological sciences. Chapter 1 focuses on the findings of farmers and plant breeders, the forebears of modern geneticists who laid the groundwork for the scientific revolution that followed. The extraordinary power of modern genetic research derives from a blend of classical approaches, which are ideally suited to unbiased explorations of biological processes, and molecular techniques that provide biochemical explanations for the principles of inheritance and cellular function. In chapters 2 and 3, each of these analytical approaches is presented in the chronological context of its historical development, providing a colorful account of the discoveries that led up to our current quantitative understanding of organic life.

Armed with these abstract concepts, the student can proceed to chapters 4 and 5, which describe the rise of genetic engineering and its newly hatched offspring, genome research. Over the last 20 years, scientists have gained the extraordinary ability to identify and manipulate essentially any of the 25,000 human genes in each of our cells. Comparative analyses of genomes from numerous organisms carried out by hundreds of sequencing centers all over the world are yielding unexpected insights into our evolutionary ancestry. Areas of the genome previously considered to be information deserts have now been revisited with the discovery of non-protein-coding RNAs, underscoring the utility of anything and everything that is maintained over evolutionary time. Comparative human genomics has also heralded a new practice of medicine, in which inherited disorders have assumed increased significance.

The rapid progress in genetic research has already had a considerable impact on philosophy, ethics, law, and religion. The prominent role that genetics will have in the evolution of society and its effect on our way of life is the subject of chapter 6, in which ethical issues arising from the genetic revolution are discussed. In the medical world, the practical repercussions of having access to individual genome sequence information are huge, as evidenced by the heavy investments by large pharmaceutical companies in medical genomics, from which they are hoping to create new drugs, new diagnostic tools, and new gene-based therapies. Because genetics will increasingly generate important public policy issues, including the proper uses of and access to human sequence data, it is critical that society be sufficiently educated to be able to make informed decisions.

Despite these caveats, what comes through in this account of society's ongoing transformation through genetic engineering is a sense of optimism. This hopefulness for a better world through genetic knowledge is justified by belief in the intrinsic value of the scientific method and by the extraordinary progress over the past century, bringing changes so sweeping in scope that they have modified our perceptions of life on Earth. A basic knowledge of the structure and function of genetic material is essential to an understanding of most aspects of a living organism. Through engaging accounts of the men and women who drove the discoveries, *Genetic Engineering* presents this basic knowledge in an unforgettable way. As Thomas Kuhn remarked, revolutions in science occur "not despite the fact that scientists are human but because they are." Understanding science as it evolves is a uniquely human pursuit and critical for the proper evaluation of the risks linked to the discoveries and revelations of this increasingly powerful field of biotechnology.

—Nadia Rosenthal, Ph.D.
Head of Outstation
European Molecular Biology Laboratory
Rome, Italy

Preface

In laboratories, clinics, and companies around the world, an amazing revolution is taking place in our understanding of life. It will dramatically change the way medicine is practiced and have other effects on nearly everyone alive today. This revolution makes the news nearly every day, but the headlines often seem mysterious and scary. Discoveries are being made at such a dizzying pace that even scientists, let alone the public, can barely keep up.

The six-volume Genetics and Evolution set aims to explain what is happening in biological research and put things into perspective for high-school students and the general public. The themes are the main fields of current research devoted to four volumes: *Evolution, The Molecules of Life, Genetic Engineering,* and *Developmental Biology.* A fifth volume is devoted to *Human Genetics,* and the sixth, *The Future of Genetics,* takes a look at how these sciences are likely to shape science and society in the future. The books aim to fill an important need by connecting the history of scientific ideas and methods to their impact on today's research. *Evolution,* for example, begins by explaining why a new theory of life was necessary in the 19th century. It goes on to show how the theory is helping create new animal models of human diseases and is shedding light on the genomes of humans, other animals, and plants.

Most of what is happening in the life sciences today can be traced back to a series of discoveries made in the mid-19th century. Evolution, cell biology, heredity, chemistry, embryology, and modern medicine were born during that era. At first these fields approached life from different points of view, using different methods. But they have steadily grown closer, and today they are all coming together in a view of life that stretches from single molecules to whole organisms, complex interactions between species, and the environment.

The meeting point of these traditions is the cell. Over the last 50 years biochemists have learned how DNA, RNA, and proteins carry out a complex dialogue with the environment to manage the cell's daily business and to build complex organisms. Medicine is also focusing on cells: Bacteria and viruses cause damage by invading cells and disrupting what is going on inside. Other diseases—such as cancer or Alzheimer's disease—arise from inherent defects in cells that we may soon learn to repair.

This is a change in orientation. Modern medicine arose when scientists learned to fight some of the worst infectious diseases with vaccines and drugs. This strategy has not worked with AIDS, malaria, and a range of other diseases because of their complexity and the way they infiltrate processes in cells. Curing such infectious diseases, cancer, and the health problems that arise from defective genes will require a new type of medicine based on a thorough understanding of how cells work and the development of new methods to manipulate what happens inside them.

Today's research is painting a picture of life that is much richer and more complex than anyone imagined just a few decades ago. Modern science has given us new insights into human nature that bring along a great many questions and many new responsibilities. Discoveries are being made at an amazing pace, but they usually concern tiny details of biochemistry or the functions of networks of molecules within cells that are hard to explain in headlines or short newspaper articles. So the communication gap between the worlds of research, schools, and the public is widening at the worst possible time. In the near future young people will be called on to make decisions—large political ones and very personal ones—about how science is practiced and how its findings are applied. Should there be limits on research into stem cells or other types of human cells? What kinds of diagnostic tests should be performed on embryos or children? How should information about a person's genes be used? How can privacy be protected in an age when everyone carries a readout of his or her personal genome on a memory card? These questions will be difficult to answer, and decisions should not be made without a good understanding of the issues.

I was largely unaware of this amazing scientific revolution until 12 years ago, when I was hired to create a public information office at one of the world's most renowned research laboratories. Since that time I have had the great privilege of working alongside some of today's greatest researchers, talking to them daily, writing about their work, and picking their brains about the world that today's science is creating. These books aim to share those experiences with the young people who will shape tomorrow's science and live in the world that it makes possible.

Acknowledgments

This book would not have been possible without the help of many people. First I want to thank the dozens of scientists with whom I have worked over the past 12 years, who have spent a great amount of time introducing me to the world of molecular biology. In particular, I thank Volker Wiersdorff, Patricia Kahn, Eric Karsenti, Thomas Graf, Nadia Rosenthal, and Walter Birchmeier. My agent, Jodie Rhodes, was instrumental in planning and launching the project. Frank Darmstadt, executive editor, kept things on track and made great contributions to the quality of the text. Sincere thanks go as well to the production and art departments for their invaluable contributions. I am very grateful to Beth Oakes for locating the photographs for the entire set. Finally, I thank my family for all their support. That begins with my parents, Ed and Jo Hodge, who somehow figured out how to raise a young writer, and extends to my wife and children, Gabi, Jesper, Sharon, and Lisa, who are still learning how to live with one.

Introduction

An incredible revolution is happening, one that has already begun to change society and will probably affect nearly everyone alive today in dramatic ways. For the past 150 years, it has been going on quietly, in the test tubes and cell cultures of research laboratories. But a glance at any day's headlines or the food labels in grocery stores reveals that this revolution is spreading far beyond the laboratory. Scientists are learning to manipulate the genes of plants and animals, changing existing organisms and creating new ones. By taking control of the mechanisms that govern heredity in other species, humans are actively directing their own evolution. Breakthroughs have come so quickly that many people find them strange and frightening. The aim of *Genetic Engineering* is to tell the fascinating story of where genetics came from, what it has made possible, and where it is likely to take humankind. It is intended for high school students and a much broader audience with little or no scientific background.

The scope of the story can be captured by two events that happened within 50 years and about 100 miles (160 km) of each other. The first occurred in the valley of Tehuacán, in southern Mexico, a rugged region of stones, dust storms, and dried riverbeds. Today the area does not look like the cradle of an agricultural revolution, but when archaeologist Richard MacNeish arrived in 1960, he was tracking one of the world's most important crops, maize, back to its ancient origins. In the cool floors of the caves of Tehuacán, MacNeish and his team found traces of thousands of years of agriculture, during which Native Americans transformed a plant called *teosinte* into tall cornstalks. It was a prime example of how ancient people across the world altered native species into the crops that now feed us.

The second event occurred in 2004 during a scientific conference in Mexico City on genetics, and it once again concerned maize. When researchers arrived at their hotel, 300 angry dem-

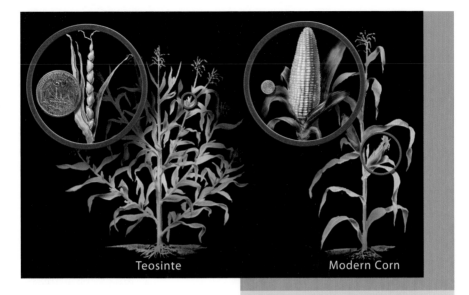

Teosinte   Modern Corn

onstrators greeted them. The protest was over a new scientific study claiming that genetically modified corn from the United States, banned in Mexico, had jumped the border and was contaminating Mexican crops. Later studies

Thousands of years of breeding produced modern corn (right) from the ancient wild plant *teosinte* (left). Modern studies reveal that changes in just five genes were responsible for the transformation. *(Nicole Rager Fuller, National Science Foundation)*

showed that this was probably a false alarm, but the incident was a sign that ecologists and farmers across the world have become concerned that native "natural" plants will one day be replaced by "artificial" species created by science. Others see these new crops as a great hope—possibly the only hope—for a world whose population is expanding much faster than it can be fed.

Genetic science began as an attempt to understand heredity and the breeding practices that allowed people to alter plants and animals for thousands of years. Today it is a science that permits the addition or removal of genes from an organism, transplantations of genes between species, and even the creation of artificial organisms. From one point of view, this is the promising continuation of a long tradition of farming and breeding. From another, it seems frightening and full of risks. Both perspectives are important to consider as society learns to cope with the "brave new world" that genetic science is creating.

Genetic Engineering traces the history of genetic science up to the present day and proposes some thoughts about how it is likely to affect the future. The first chapter recounts how early breeding and agriculture were transformed into a science through a series of studies carried out by the monk Gregor Mendel, whose work lay forgotten in libraries for nearly 40 years. Its rediscovery launched classical genetics, described in the second chapter, a period in which scientists brought flies and other organisms into the laboratory in hopes of discovering new genes and learning how they function. It quickly became clear that to succeed, scientists would need to know what genes were made of. This was finally achieved through a series of breakthroughs in chemistry and physics that culminated in an explanation of the double-helix structure of DNA. The discoveries ushered in a new era known as molecular genetics (chapter 3), during which scientists began to understand how genes guide the activity of cells and organisms.

The 1970s and '80s saw the development of a set of tools that allowed scientists to "read and write" in the language of genes (described in chapter 4). Almost immediately, these methods led to applications in medicine and agriculture. The last two chapters describe some of the developments in the first few years of the 21st century and how society is coping with some of the ethical challenges that accompany these changes.

Politicians and the public now face tremendously important choices about how genetic engineering and its products should be used. These decisions will influence the lives of people across the world. Understanding the science behind the modification of crops, new types of medicine, and other products will not give a single definitive answer to questions such as whether it is a good idea to grow a new type of corn in Africa, or whether mosquitoes that are immune to malaria should be released into the environment. But it can help put things into perspective and give some idea of the risks involved. With so much potentially at stake, everyone should have a basic grasp of the science behind the issues.

1

From Breeding to a Science of Heredity

Off the coast of Japan lives a species of crab that is said to be possessed by the ghosts of warriors. The legend dates back to the Battle of Dannoura in 1185, when a group of samurai was unable to defend the seven-year-old emperor, Antoku. Upon his death in the fight, the conquering army threw Antoku's warriors into the sea. According to tradition, their ghosts roam the murky depths nearby, and today, when fishermen in these waters pull in their nets, they find crabs whose shells look remarkably like the faces of ferocious samurai. The British biologist Julian Huxley (1887–1975) said that humans are the cause of this strange phenomenon. Centuries ago, some of the crabs had shells that slightly resembled faces, which led the fishermen to throw them back. Each generation, the resemblance grew stronger as fishermen threw back the animals that most resembled samurai. Today every Heike crab bears the famous scowl.

Throughout history, humans have shaped other living organisms, sometimes intentionally, sometimes accidentally. They have improved the plants and animals that provide food or selected them for other reasons, including purely aesthetic ones. Heike crabs are one example; the Chinese did something similar with their dogs, selecting pets that resembled the spirit lion of Buddha to produce Pekingese dogs. Every species alters its neighbors (and is altered by them); humans are the only species that does so deliberately. This process has been going on since ancient times, when farmers and breeders observed that plants and animals inherited the characteristics of their parents. Only in the 19th century, however, did scientists begin to systematically investigate heredity. The work of Gregor Mendel laid the foundations of a new science of genetics, which later led to powerful new methods of investigating and manipulating species. This chapter covers people's understanding of heredity up to that key moment in the history of science.

THE ORIGINS OF DOMESTIC PLANTS AND ANIMALS

A typical meal today is like a world map of prehistory. Start with salad: Lettuce was probably first cultivated from wild plants on the tiny Mediterranean island of Malta. Tomatoes arose in Central America, which was also the birthplace of corn, squash, and kidney beans. Potatoes and peanuts were domesticated from plants native to South America. The ancestors of sunflowers and strawberries probably first grew in areas that are now in the United States. Asia was the source of soybeans, apples, rice, and onions. Africa produced sorghum and coffee, and sugarcane and bananas originated along the Pacific Rim. The world's breadbasket was the "Fertile Crescent"—a region encompassing parts of ancient Egypt, Mesopotamia, and other parts of the Middle East—where wheat and rye were cultivated from grasses. Peas and grapes also came from this region.

Over thousands of years, farmers selected the most useful forms of many plants, able to provide more food or better suited for other purposes. Modern versions are usually larger, grow longer, and have been adapted to a wider range of climates than their wild ancestors. Yet some changes have made plants completely dependent on humans. Modern corn would die out quickly if it were not for farmers. Wild maize lets its seeds fall to the ground, where they grow to become the next generation, but modern corn has been bred to hold on to its kernels so they

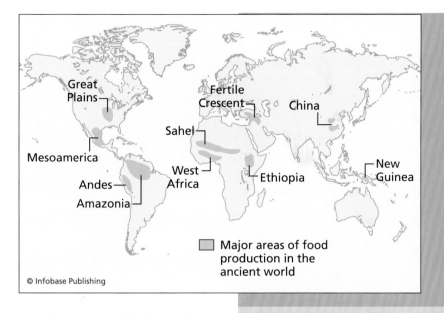

© Infobase Publishing

can be eaten. Without human help, its seeds would not survive until the next planting. Peas have undergone a similar evolution. In the wild, their pods need to explode to spread their seeds. Sometimes a *mutation* in one of the plant's *genes* leaves a pod intact. That

These regions in the prehistoric world played important roles in the development of agriculture. Here ancient people transformed wild plants into crops such as wheat, corn, and rice, which spread to become mainstays of the human diet throughout history and feed the world today.

would be useless and probably fatal in nature; on the other hand, it is a good characteristic for a food and was selected and cultivated by farmers. Today, when people eat peas, they are eating seeds that originated as underdeveloped mutants.

In other cases, a single species has been molded into many forms as farmers found different parts of the same plant desirable. Cabbage, cauliflower, kohlrabi, broccoli, and brussels sprouts are modern versions of a single ancestor. Broccoli was chosen for its stems and flowers; kohlrabi developed as farmers selected a part of the plant called the storage stem; and cauliflower is the result of selecting luscious flower clusters.

Centuries or millenia of breeding have changed most plants so dramatically that it is often difficult to identify the wild an-

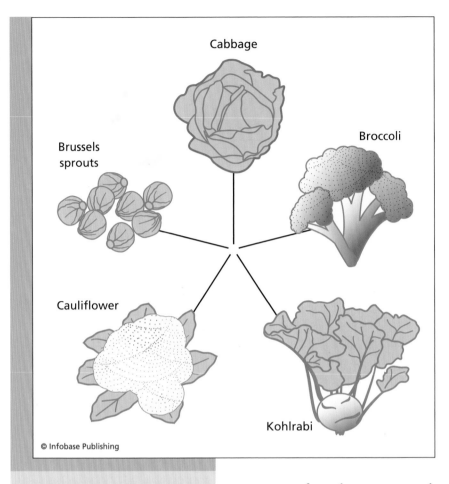

Cabbage

Broccoli

Brussels sprouts

Cauliflower

Kohlrabi

© Infobase Publishing

Sometimes farmers selected varieties of a single ancient plant for different reasons, developing it into crops that look quite different today. Clockwise from the top, cabbage, broccoli, kohlrabi, cauliflower, and brussels sprouts are descendants of one common ancestor.

cestors of modern crops such as corn. Archaeologist Richard MacNeish (1918–2001) believed that maize stemmed from a Central American plant called *teosinte,* which has dwarflike cobs. Its kernels are small and hard, but when tossed into the fire, they pop and can be eaten; they can also be ground into meal and fried on stones in flat, tortillalike forms. It took modern science and *DNA* comparisons to finally prove MacNeish right. Although the two types of plants look quite different, it took changes in only five teosinte genes to transform the wild plant into modern corn—such small differences that they are now considered to be the same species.

Comparisons of DNA are now being used to solve several similar mysteries about the origins of other foods, domestic animals, and even human populations. The recent completion of the dog genome proves that even the meekest of today's pets stems from wolves; the genome also revealed that dogs were tamed more than once, in different places. Horses also have a single common ancestor, but that species had spread widely across Asia before several human tribes independently figured out that the animal could be ridden and put to work. Cows were domesticated twice in ancient times: One wild group sired the cattle of Europe and Africa, while the holy cows of India are the descendants of animals captured and tamed elsewhere.

Most changes that produced modern species happened gradually, but hybridization—crosses between different species—could quickly produce new forms. Hybridization happens frequently in the wild, when pollen from one type of plant fertilizes another; it is also a powerful tool for farmers. By mixing the hereditary material of two species, hybridization often leads to massive genetic changes. The grains used to make bread are good examples: Wheat arose from a combination of wild einkorn with goat grasses. Einkorn wheat has 14 *chromosomes,* whereas modern bread wheat has 42; so, ancient hybridization events and cultivation produced a wheat genome with more than 10 billion *base pairs* (the single chemical "letters" that make up DNA). Scientists are only beginning to understand the impact that this has had on the plant, which likely acquired far more genes than humans (whose DNA holds three billion bases). A cross between radishes and cabbages yields offspring with double sets of chromosomes—in other words, two complete sets of genetic material—that have combined in completely unpredictable ways. These facts will be important to remember in chapters 4 and 6, which discuss genetically modified crops, their impact on the environment, and the public perception of their safety. In those cases, scientists have usually changed just a few genes and carried out careful laboratory studies of their effects on plants and other organisms.

THE RISE OF AGRICULTURE AND DOMESTICATION

Modern humans have probably existed for about 100,000 years, and for more than 90 percent of that time, they lived as hunter-gatherers, following herds of prey and living off of food that they could collect relatively easily. This way of life is still practiced by a small number of people in the Andaman Islands of the Indian Ocean and in the Great Victoria Desert of Australia, but nearly everywhere else it has died out. There are several reasons: A hunter-gatherer community requires 10 to 100 times more land to survive than a group of the same size that practices agriculture, partly because very little of what grows in the wild is edible. Being constantly on the move means that hunter-gatherers are unable to store much food or care for as many children as city dwellers.

Today's agricultural traditions probably arose in about 10,000 B.C.E. at the end of a great ice age. Climate change offered new, fertile lands but also meant that humans had to change lifestyles or follow old sources of food to new places. Populations increased rapidly during this period and soon outgrew the available resources. As more of the world became settled, it became impossible to migrate without encroaching on someone else's territory, leading to competition and wars.

Carl Sauer (1889–1975), a geographer at the University of California, Berkeley, and Lewis Binford (1930–), an archaeologist at Southern Methodist University in Texas, proposed that a key motivation for the invention of agriculture was need, as there was no other way to support large populations. Chance may also have played an important role. Some regions were so abundant in plants and animals that they could support large settlements even without organized agriculture. These towns would have had waste dumps, whose fertile soil created ideal growing conditions for any seeds that were thrown out with the trash. Thus agriculture might have begun as harvesting at the town dump and then spread as cultures came into contact with each other.

Settling in one place gave people time to observe local plants and animals and to learn new ways of using them. Even very early sites reveal a rich use of plants. Archaeological studies of a settlement called the Tell Abu Hureyra in Syria, founded around 11,500 B.C.E., yielded the burned remains of 157 different species of plants collected by its inhabitants. Most had seeds that were edible or could be eaten after simple treatments to remove their poisons. Others would have been useful as medicines or dyes.

But this intimate knowledge of plants and animals stretches back farther into history. Hunter-gatherers, too, develop a very sophisticated familiarity with the species they depend on. "Bio-geographer" Jared Diamond (1937–), a professor at UCLA, recounts that the native New Guineans with whom he lived for many years can identify 29 edible mushrooms without mistaking them for similar poisonous types. On meeting strangers, hunters collect information instead of souvenirs, asking those they meet about local species.

By about 10,000 B.C.E., people in many regions were systematically cultivating plants. Animals seem to have been domesticated even earlier. Dogs accompanied bands of hunter-gatherers as long ago as 15,000 B.C.E. Pigs, sheep, goats, and silkworms were raised in the Fertile Crescent and Asia. Rock paintings from South Africa and India show people hunting for honey; bee-keeping was practiced in many regions. By 8500 B.C.E., people in the Fertile Crescent had become very active food producers. They domesticated wheat, peas, and olives. A thousand years later, groups in parts of present day China were growing rice and millet. Agriculture was probably invented independently in the Americas and Africa a few thousand years after that. The founder crops developed in these places were so valuable that they quickly spread to other parts of the world.

In his book *Guns, Germs, and Steel,* Diamond proposes that differences in the plants and animals available for cultivation, combined with climate and geographical factors, created advantageous conditions for a few prehistoric groups. The Fertile Crescent's lush valleys tucked between mountains offered farmers a wide variety of wild plants to choose from and allowed settlements to take advantage of longer growing seasons; the same plants could be cultivated at different times by raising them at different altitudes. The east-west geography of Europe and Asia created huge areas with similar seasons and climates where the same crops could be grown. In the Americas, which stretch north to south, this was much harder. Diamond believes that such factors contributed to the Europeans' conquest of much of the rest of the world. Cultivation pushed the development of technology, as farmers invented tools to harvest their crops more quickly. Sticks were used as crude plows in central Europe in about 5000 B.C.E. A thousand years later, people began to hook them to oxen, allowing them to farm in areas where the ground was harder.

Agriculture prompted enormous changes in human society. Tending fields meant living nearby and guarding them, which encouraged the development of villages, governments, and religions. Obtaining food required less time, so a society could support lawgivers, priests, soldiers, craftsmen, artists, and other specialists. This created more complex social structures, and it gave cities objects for trade.

Farming practices became deeply tied to religious rituals and the beginnings of science. Planting the same crops every year gave people the chance to observe the seasons, helped them discover the best times for planting and harvesting, and undoubtedly prompted the development of calendars. Careful observation of plants and animals revealed that "like begets like"—plants and animals tend to have offspring that resemble their parents—the first step toward taking control of heredity and nature. Farmers learned to sow seeds from the most productive plants of the year before and to control the mating of their animals.

Not all the effects of farming were positive. Clearing land for fields created breeding grounds for mosquitoes, which carried parasites between humans and their domestic animals and led to new forms of disease. Tuberculosis, smallpox, the flu, and many other diseases first appeared in animals but evolved into forms that could infect people. Europe and Asia had more types of wild animals suited for domestication, which meant more diseases. This had a striking influence on human evolution because people with partial resistance to parasites were likely to have more children. The best example is malaria, caused by a one-celled organism and carried by mosquitoes. With the rise of farming, it became a significant threat to humankind. However, people suffering from a genetic blood disease called *sickle-cell anemia* have some protection from malaria, so sickle-cell anemia is found at much higher rates among human populations where the incidence of malaria is also high.

EARLY IDEAS OF INHERITANCE

As ancient people became more dependent on crops and domestic animals, they began to observe other species carefully and to encode rituals of breeding and agriculture in traditions, religions, and law. Genesis, the first book of the Bible, tells how Jacob was given the job of tending his father-in-law's flock of goats. He was allowed to keep goats with spotted coats but had to turn those of pure color over to his father-in-law. In the story, Jacob used magic rituals to increase his share of the herd, but clever breeding could have accomplished the same thing. By allowing the right animals to mate, genes for spotted fleece could spread through the flock.

Under Jewish law, baby boys were required to be circumcised, but exceptions were permitted when *hemophilia* was known to run in a family. This reflects an appreciation for patterns of inheritance. Hemophilia arises because of a defect in a molecule that helps blood to clot. The disease is often fatal because slight injuries can cause unstoppable bleeding. Rabbis ruled that if a baby's uncle (on the mother's side) had died during circumcision, the ceremony did not have to be performed. This means that hemophilia was recognized to be hereditary and somehow connected to the mother's side of the family. Today scientists know why: The disease is caused by a defective gene on the *X chromosome,* which males always inherit from their mothers.

But despite thousands of years of farming and domestication, there was still no deep understanding of heredity until modern

The Girl Who Gave Birth to Rabbits

The strange story of Mary Toft was first reported in a sensationalist newspaper in England called *Mist's Weekly Journal,* named after its owner and printer, Nathaniel Mist. The paper was so critical of the government that a spy was hired to keep an eye on Mist's activities. This spy was none other than Daniel Defoe, who wrote his novel *Robinson Crusoe* while working at the paper.

On November 19, 1726, the journal printed an article about a "strange, but well attested piece of news" from the village of Godalming, about 20 miles (36 km) southwest of central London. John Howard, a doctor and midwife, arrived at the home of a young pregnant woman named Mary Toft whose baby was about a month overdue. Howard delivered the baby, which turned out to be "a creature resembling a rabbit; but whose heart and lungs grew without its belly." The story goes on:

> About 14 days since she was delivered by the same person of a perfect rabbit; and, in a few days after, of 4 more; and on Friday, Saturday, and Sunday, the 4th, 5th, and 6th instant, of one in each day; in all nine. They died all in bringing into the world. The woman hath made oath, That two months ago, being working in a field with other women, they put up a rabbit; who running from them, they pursued it, but to no purpose: This created in her such a longing to it, that she (being with child) was taken ill, and miscarried; and, from that time, she hath not been able to avoid thinking of rabbits.

times, as revealed by the curious tale of the "rabbit woman" of the early 18th century.

Explaining why humans could not give birth to rabbits and other questions about heredity involved two great mysteries.

Folk customs of the time held that experiences during pregnancy could influence the health of a child, but for a woman to give birth to rabbits stretched the belief of even the most credulous readers. King George I ordered an investigation, and Mary Toft and her doctor were brought to London. On December 1, she seemed to be going into labor again, but she ate a large dinner including, ironically, a dish of rabbit. "Great numbers of the Nobility have been to see her," reported the *London Journal*, on December 3, "and many Physicians have attended her, in order to make a strict Search into the Affair; another Birth being soon expected."

December 4 marked the beginning of the end of Mary's brief burst of fame. A porter at the prison where she was being held testified that she asked him to obtain a rabbit for her from the market. Mary denied having done so, but her sister, who was helping to take care of her, said that they had asked for a rabbit—but only to eat.

The judge in the case was losing patience and threatened to subject her to an "extremely painful procedure" if she refused to tell the truth. Finally, on December 7, she began her confession. She claimed that after a recent miscarriage, a female friend had encouraged her to pretend to give birth to rabbits—a scheme that would make them rich. The accomplice had provided the rabbits and helped her insert them in her birth canal. Mary Toft was charged with fraud, was brought before the courts, but was released without punishment, perhaps out of pity for a woman so desperate for a moment of fame.

First, what physical material was involved in transmitting characteristics from parents to their offspring? It surely involved the fluids exchanged by animals during sex, but beyond that, scientists were mystified. In today's information age, people are used to objects that change form. A voice can be transformed into electrical impulses and transmitted across the world; movies can be captured as tiny pits burned onto the surface of DVDs, so it is not so hard to imagine an organism's building plan stored as a set of chemical instructions inside of cells. Today it is also common knowledge that everything is composed of molecules and atoms that can be mixed and recombined in different ways to create substances with different properties. But this "atomistic" view was not accepted by scientists until the 19th century, when chemists began to learn to break down substances and gases.

The second mystery had to do with the way embryos grew and took form, seemingly out of nowhere. A seed did not look anything like a tree, but it could become one, and a tiny ball of flesh could develop into an entire human being. This phenomenon was as magical as if applesauce were suddenly to transform itself into apples. Several explanations were offered. The ancient Greek physician Hippocrates (ca. 460–370 B.C.E.), known as the "father of medicine," proposed a hypothesis about heredity called *pangenesis.* He suggested that "particles of inheritance" arose in all parts of the mature body of the parent and were collected into the fluids exchanged during sex. The idea was attractive because it meant that information about body parts was somehow stored in particles that could be directly transmitted to the new organs that arose in embryos. A sort of struggle between the fluids of the father and the mother determined which features were inherited from which parent. Assuming that both parents contributed equally to an offspring was a promising beginning; unfortunately, Hippocrates' ideas were supplanted by those of a more influential philosopher.

Aristotle (384–322 B.C.E.) claimed that a father's semen played the dominant role in heredity. The mother provided an embryo's raw material, he wrote, but its structure and features came from the father. This was incorrect, of course, but it contained an important grain of truth: Heredity involved a hidden plan by which unformed material takes shape. This concept, too, would be practically forgotten until modern times. Aristotle was a sharp observer and pointed out other important clues about the hidden plan: Inherited features did not appear in a child all at once, at birth. Baldness, gray hair, or complex behaviors could clearly be inherited, although they appeared only late in life. They might even skip over a generation and reappear in grandchildren.

Like many others, Aristotle wrongly assumed that some experiences and characteristics that an organism learned or acquired over its lifetime could be transmitted through heredity—for instance, supposing that Arnold Schwarzenegger's children would be born with the physique he developed through training. Yet the philosopher recognized that other traits, such as the loss of an arm or leg, could not be. Aristotle discussed hybrids such as the mule and proposed that many other animals might have arisen as crosses between different species. Giraffes, for example, might be a cross between leopards and camels. He assumed that this could happen only between animals that were somewhat similar to each other; later authors were not so cautious. (One even proposed that the ostrich might be a hybrid of a sparrow and a camel, so its scientific name, *Struthio camelus*, comes from the Greek for "sparrow-camel.")

Heredity was thought to work differently in plants and animals until scientists discovered that plants reproduce sexually. The Englishman Nehemiah Grew (1641–1712) recognized that pollen played the role of the male element in the fertilization of plants. At the end of the 17th century, the German Rudolph Camerarius (1665–1721) identified the pollen-bearing sex organs in males, called anthers, and structures in females called stigmas to which the pollen attaches itself. From there, the pollen penetrates to the plant ovary (for example, the pods of peas), where seeds form.

Pollen from the pistils of one plant usually has to reach the female sex organs of another to produce fertile seeds. Camerarius proved this by showing that if he cut off the anthers of male plants, the females would not reproduce. Two decades later, Carl Linnaeus (1707–78) of Sweden repeated the experiments and showed that such sterile plants once again produced seeds if he did the pollination himself. Linnaeus then tried crossbreeding experiments in which he applied the pollen from one species to another, sometimes obtaining hybrid species.

Understanding fertilization in mammals lagged behind because scientists had not yet discovered the equivalent of pollen or eggs. Antoni van Leeuwenhoek (1632–1723) discovered sperm under the microscope. While some researchers believed the cells to be parasites, Leeuwenhoek claimed he had found the male reproductive material and suggested that a single sperm was enough to fertilize an egg. He also proposed that each sperm contained a complete preformed organism—folded up into a tiny space. Later writers carried this idea to a fantastic extreme, hypothesizing that tiny sperm contained even tinier sperm of the following generation, on and on, from the beginning of humanity to the last generation. This confused the issue of heredity because there seemed to be no way a mother could contribute to a child's hereditary material.

One hypothesis explaining how a large, complex organism could arise from a tiny embryo was that sperm contained a miniature human body, already containing most adult features, folded up very tightly. *(Andrew Canessa)*

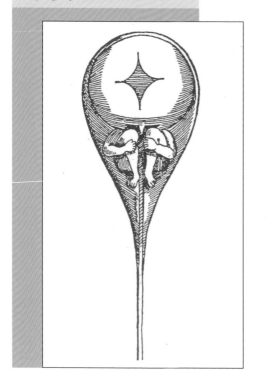

The 18th century saw the beginning of scientific studies of human heredity. Pierre-Louis Maupertuis (1698–1759) tracked the appearance of extra fingers through four generations of a family—the first known description of a genetic disorder in humans; he also constructed family trees of albinos and investigated color patterns in dogs. A half century later, Joseph Adams (1756–1818) wrote *A Treatise on the Supposed Hereditary Properties of Diseases,* in which he recognized that mating between close relatives frequently led to disorders. He claimed that some hereditary diseases became apparent only late in life and that the environment often played a role in how they developed.

In the early 1800s, major improvements in microscopes suddenly gave scientists a much sharper view of the cellular world, allowing them to observe fertilization in plants directly for the first time. Giovanni Amici (1786–1863) showed that the ovaries of orchids contained a single cell that remained inactive until pollen arrived, then it was fertilized and developed into a seed. William Harvey (1578–1657), studying chickens, had already made a similar claim for animals—that they all originated as eggs. But the fact that mammals produce very few eggs (women produce only one a month) made them hard to find. Finally, Karl Ernst von Baer (1792–1876) proved Harvey right in 1827, when he identified an egg cell within a dog. The cells that carried hereditary information had been identified; now the challenge was to discover the form of the information itself.

HEREDITY AND THE THEORY OF EVOLUTION

A science of heredity needs to explain why humans give birth only to humans, rather than to rabbits, but also why babies are not perfect copies of their parents. Both the accuracy and the imperfections of heredity are central concepts in the theory of evolution, announced to the world in 1858 by Charles Darwin (1809–82) and Alfred Wallace (1823–1913). The fact that not all members of a species are identical is called variation, and small differences give some individuals a better chance of reproducing and passing along their hereditary material. They may be more fertile than others, or they may be more likely to live long enough to reproduce. So, variation results in natural selection, which explains how the balance of traits in a species changes over time and eventually yields new species.

A photo of Charles Darwin taken by Julia Cameron in 1869, 10 years after the publication of *On the Origin of Species*. Heredity was a cornerstone of evolution. To be compatible with the theory, genetics would have to explain how organisms could pass characteristics to their offspring, and how mistakes in the process could occur. *(University of Brunel)*

By the 1840s, Darwin had grasped how these factors work together to form new species, but he waited nearly 20 years to publish his ideas, intending to support his radical new theory with as many facts as possible. He began an exhaustive study of breeding, supposing that farmers would best understand heredity because they had been manipulating species throughout history. They had achieved amazing things: creating sheep with short legs, hornless cattle, and an amazing variety of pigeons. Darwin gathered as many types of pigeons as possible and carefully studied their anatomy. Any expert who did not know their history, he wrote, would surely claim they were different species. Instead, they all descended from a common ancestor, which Darwin believed to be the rock pigeon.

The entire first chapter of *On the Origin of Species* is devoted to past work on the subject and Darwin's own experiments. It dismisses another popular "evolutionary" theory of the day, proposed by Jean-Baptiste Lamarck (1744–1829), who claimed that lifetime experiences and changes in an organism could be inherited by its offspring. This was a "use it or lose it" philosophy, which supposed that living in the dark would cause species to lose their eyes, or that parents who developed huge muscles through exercise would pass them along to their children. The best-known example was his hypothesis that giraffes developed long necks because, over many generations, they stretched to reach leaves in the upper branches of trees.

Darwin believed that an organism's hereditary material was set in place before its birth—otherwise, all the newborns in a litter of cats or dogs would be identical. He also noticed some "whimsical" correlations between inherited characteristics: Blue-eyed cats were always deaf, hairless dogs had bad teeth; and pigeons with big feet came with long beaks. Half a century later, an explanation would be offered: Genes located on the same chromosome are usually inherited together. Careful research led Darwin to other deductions that have only been proven in the age of genomes. By comparing DNA sequences from many different organisms, scientists have confirmed his hypotheses that the greyhound, bloodhound, terrier, spaniel, and bulldog descend from a single, wolflike ancestor, that the chimpanzee is the closest relative of humans, and that *Homo sapiens* probably originated in Africa.

While Darwin's ideas about the effects of heredity were correct, he was wrong about the mechanisms by which it worked. He believed in a pangenesis hypothesis, like that of Hippocrates, mentioned in the previous section. Darwin proposed that the hereditary material of animals was transmitted in the fluids exchanged during sex. Organisms possessed reproductive "particles" called *gemmules,* normally dispersed throughout the body and then collected in the sex organs. These mingled during sex to create new organisms that blended characteristics from both parents. Other scientists were skeptical, recognizing that such blending would "dilute" the effects of inheritance and make it impossible for favorable traits to gain a foothold in a species.

Darwin's cousin Francis Galton (1822–1911), famous today for his role in developing fingerprints as a system for use by the police, hoped to prove the pangenesis hypothesis in a series of experiments. Gemmules had to flow through the blood, he reasoned, so he transfused blood from rabbits of different colors into a silver-gray strain. He expected that this would produce offspring of mixed colors. They all remained silver-gray, convincing most scientists that the hypothesis was wrong. Darwin replied that gemmules did not necessarily have to be transmitted in the blood and clung to the pangenesis idea.

This and similar ideas have been called "analog" hypotheses of heredity, in contrast to the "digital" view developed later by Mendel and the early geneticists. Darwin thought that organisms' characteristics lay along a smooth scale—people are short to tall, heavyset or skinny, or anywhere in between. He failed to realize that many characteristics were transmitted in units (genes)—and followed an "either/or" pattern. Thus, peas that inherit one form of a gene are wrinkled; another form of the same gene makes them round. A middle form—round with a few wrinkles—might not exist at all.

One of the most important aspects of evolutionary theory was to connect variation within a species to differences between species. The small differences produced by heredity could produce much bigger differences when natural selection worked on a species over millions of years. But proving that this was the case required an accurate understanding of heredity.

GREGOR MENDEL DISCOVERS THE LAWS OF HEREDITY

As Darwin was finishing *On the Origin of Species,* the pieces of a scientific theory of heredity were finally coming together in a monastery on the European mainland. Today this might seem strange, but in the 19th century, monasteries and abbeys offered an education and career for promising students from poor families. This was the situation of Gregor Johann Mendel (1822–84), who grew up on a farm north of Moravia (today part of the Czech Republic). Mendel had frail health and was clearly unsuited for the hard life of the farm. He was a bright pupil, and his sister used part of her inheritance to help send him to the University of Olmütz, where he studied mathematics and physics. He hoped to become a teacher, but a nervous condition caused breakdowns when he had to take tests.

An abbey seemed to be the only choice left. His physics professor advised him to enter St. Thomas in the town of Brno, a famous center of learning run by one of the most liberal and relaxed orders of Catholicism, the Augustinians. (The professor, also a priest, had lived at the abbey for 20 years.) Upon becoming a monk in 1843, Mendel was given the name Gregor and used it for the rest of his life.

The abbey paid his way as he continued his physics studies at the University of Vienna. He also became interested in statistics, which would play a crucial role in his studies of heredity. Yet a piece of bad luck kept him from finishing his degree: His exams fell in the hands of a professor who disagreed with some of his "modern" scientific views.

After failing as a farmer, a university student, and a teacher, Gregor Mendel found a home at the Abbey of St. Thomas, where he quietly and systematically worked out the basic principles underlying heredity. (MendelWeb)

For example, Mendel claimed that both parents contributed equally to the heredity of their offspring. Many scientists had come to believe this, but as Mendel discovered, a single conservative professor could block a student's career. Mendel returned to the abbey, where his troubles continued. First he assumed the duties of a parish priest, but dealing with the sick and needy caused another strain on his nerves and sent him to bed for weeks.

Mendel might never have fit in had it not been for the fact that the abbey had a progressive, sympathetic abbot who relieved him of his responsibilities and gave him teaching jobs. St. Thomas was headed by Cyrill Napp, a scientifically minded man who carried out breeding experiments on sheep as a hobby. Napp took a liking to the young monk and encouraged him to pursue his own scientific investigations. Mendel began keeping mice in his room, crossing albinos with wild mice in hopes of understanding the heredity of color. The experiments never got anywhere because of church politics. Increasingly, the focus of monasteries was shifting toward intellectual pursuits. The Catholic Church felt that things were getting out of hand and began a round of inspections. Some institutions were told bluntly to go back to their original religious missions; others were shut down. When St. Thomas came under scrutiny, the visiting bishop was shocked to find animals mating in a monk's room. As part of the compromise made to keep the abbey open, Mendel's mice had to go.

What looked like yet another piece of bad luck was actually a turning point for the young monk. Abbot Napp had already given him a plot in the abbey's experimental garden, where he could study breeding in plants. No one objected to this because, as Mendel remarked, "The bishop did not understand that plants also have sex." He had decided to tackle the question of heredity in a rigorous, scientific way. He sent off for 34 varieties of peas, which he began cultivating in the monastery garden, and eventually settled on 22 strains that he would study more intensively.

Mendel achieved a breakthrough where Darwin and others had failed because he took a different and much more systematic approach to the problem of heredity. First, rather than trying to understand the plant as a whole, he focused on how single features of plants were transmitted between parent and offspring. Secondly, he controlled his experiments extremely closely, eliminating every source of contamination that he could imagine. His background in mathematics and statistics gave him the skills to discover the patterns that guided what he was seeing. Finally, he followed the fates of plants for many generations, rather than expecting to see all the rules of heredity at work in a single generation—another error made by many other scientists.

Pea plants were ideal for studying the contribution of two parents to heredity. Most plants reproduce through movement of pollen from the male sex organ, the anther, to the stigma of a female. A plant may pollinate itself, or there may be a transfer from one to another by bees, moths, or other insects.

In peas, the male and female organs lie within an inner layer of the flower called the keel, and the entire process takes place there. So under normal circumstances in the wild, the plant fertilizes itself; a single plant serves as father and mother to new seeds—the peas. It was possible to create peas from different parents, however, using the method developed by Camerarius and Linnaeus. Mendel carefully cut open the keel of one plant and removed the anthers with a pair of tweezers. He collected their yellowish pollen on a brush and moved to another plant, the "mother," which he once again cut open, and brushed the pollen onto its stigma. He had already removed the anthers of this plant, too, ensuring that the plant did not pollinate itself. He closed the keel again and wrapped it in a small bag so that it could not be reached by free-floating pollen. This also kept out flying insects, but it did not stop weevils, and, occasionally, Mendel had to eliminate some of his data because of infestations of the insects that might have caused contaminations.

He decided to study seven characteristics of peas that could be clearly identified and tracked. For example, he had one strain of plant that produced smooth, round peas. Another strain was identical, except that its seeds were wrinkled. When Mendel first obtained the plants, round-pea parents occasionally produced wrinkled offspring, and he had to breed them over and over again until they always gave the same results. The same was true of the other traits he wanted to study:

- the color of the pea (green or yellow)
- the color of the flower (purple or white)
- the color of the unripe pods (green or yellow)
- the form of the pods when they became ripe (inflated or constricted)
- the position of the flower (growing either along the stem or at its tip)
- the length of the stem (tall or dwarflike)

There were other traits he might have studied, but it was difficult to get strains that produced them reliably, and he would have his hands more than full with seven characteristics.

After two years, the strains were stable, and Mendel began crossing them. He cut open the keels of plants that produced round peas and fertilized them with pollen from the wrinkled-pea strain. He did the same thing, reversing the sexes: introducing round-pea pollen into wrinkled-pea strains. The plants produced several hundred seeds (called first-filial-generation hybrid seeds, or F1). Strangely, all of them were round. Rather than drawing any conclusions from this, Mendel waited a year and planted the F1 seeds, and this time, he allowed the plants to fertilize themselves, the natural way. The F1 plants produced 7,324 second-filial-generation peas (F2), of which 5,474 were round and 1,850 were wrinkled. This gave a proportion of 2.96:1 in the seeds of the second generation—nearly three to one. Crosses between strains with the other features gave almost identical results.

This three-to-one ratio led Mendel to several brilliant insights. He realized that each of the features he was studying (for example, wrinkled versus round) was composed of two "elements"—one inherited from each parent. One element was dominant, and the other type was recessive, meaning that if a pea inherited one of each type, it would take on the dominant form. This explained why the first-generation peas were all round: Each had inherited a round element from one parent and a wrinkled from the second. It made no difference which parent had contributed which element—the results turned out the same. This meant that the two sexes contributed equally to the characteristics of the offspring. Mendel had proved the point that had caused him to flunk his university examinations.

The principle of dominant and recessive traits also explained what happened to the second generation, when the F1 plants fertilized themselves. Each of their offspring received a chance combination of two traits. Statistics allowed Mendel to predict that one fourth of the plants had received two round elements (producing round peas); another fourth had inherited two recessive, wrinkled elements (making them wrinkled); and two-fourths had received one of each, giving them the dominant form (roundness). These proportions were carried on in later generations bred from the F2 plants. The other traits that Mendel studied followed the same pattern, which meant that they were determined also by elements with a dominant and recessive form.

For the experiments, he had assumed that characteristics were passed down independently of each other—in other words, a pea's color had no influence on whether it was wrinkled. But was that really the case? Mendel wondered what would happen if he mixed traits in a single plant, so he launched a more complicated study. He began by creating one strain of plants whose peas were always yellow and round (both dominant) and another with green, wrinkled peas (both recessive).

When he mated these plants, all the seeds in the F1 generation were yellow and round, as he expected. He anxiously waited to see what would happen with the second generation. Of the 556 peas he obtained, 315 were yellow and round, 101 were yellow and wrinkled, 108 were green and round, and 32 were green and wrinkled. These results, a ratio of about 9:3:3:1, made perfect sense if color and shape were separate units (in other words, if they were passed along from the parents independently of each other). One in every 16 peas would inherit both recessive traits. Another 16th would inherit both dominant traits; the rest would have a mixture of dominant and recessive genes.

In 1865, after 11 years of examining tens of thousands of plants and hundreds of thousands of peas, Mendel mounted the podium at the Society for the Study of Natural Sciences in Brno to present his results in two lectures. The society's members included university scientists, teachers, and other science lovers. By this time, he had extended his work to other plants, including beans, and obtained the same results. He had also read *On the Origin of Species* and thought about it carefully. His experiments said nothing about how species might adapt; in fact, they tended to support a contrary view. What looked like variation within species might be nothing more than a reshuffling of existing characteristics.

Mendel's audience appeared interested, but they were unable to judge the importance of what they were hearing. When his paper, entitled "Experiments on Plant Hybridization," was published in the *Proceedings of the Brno Society for the Study of Natural Sciences,* it received about the same response. Although

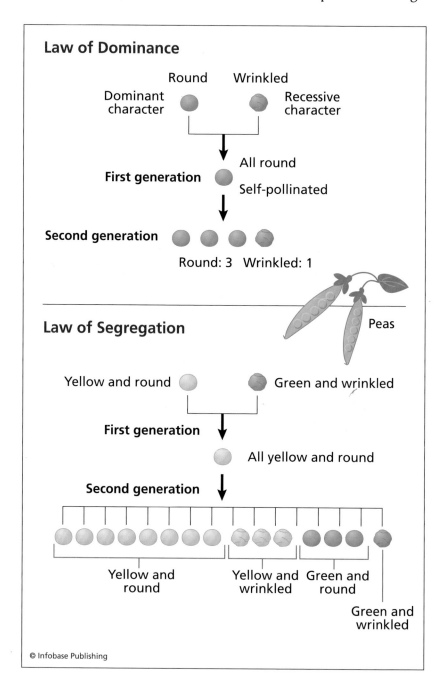

Law of Dominance

Round   Wrinkled

Dominant character   Recessive character

First generation   All round

Self-pollinated

Second generation

Round: 3   Wrinkled: 1

Peas

Law of Segregation

Yellow and round   Green and wrinkled

First generation

All yellow and round

Second generation

Yellow and round   Yellow and wrinkled   Green and round

Green and wrinkled

copies were sent to at least 120 libraries in Europe, scientists did not understand that Mendel had begun to unravel the ancient mystery of heredity.

One of the journals landed on the desk of Karl von Nägeli, a professor at the University of Munich. Franz Unger, Mendel's botany professor in Vienna, had praised Nägeli as a genius. Mendel sent Nägeli a copy of the journal with a letter. It was important to have the results verified, and Nägeli might be the right person to do so. Mendel had already been working with a plant called hawkweed, which Nägeli had experimented with, and mentioned this in the letter.

What happened then is one of the great tragedies of science. Nägeli responded critically, stating that the results were interesting but certainly insufficient to justify an entirely new theory of heredity. (Nägeli was pursuing his own hypothesis, that organisms inherit a bit of information from each parent and then blend it into an intermediate form.) Several letters were exchanged; Mendel offered to assist the professor without pay. Nägeli sent back hawkweed seeds of his own and encouraged Mendel to raise them.

The work was extremely difficult because hawkweed stamens are so tiny and so close to the pistil that they could be separated only painstakingly, under the microscope. Even when Mendel managed, the attempt was useless because of an unusual property of hawkweed. In some cases, the plant reproduces parthenogenically—a type of cloning in which only the genetic material of the mother is used to create offspring. Pollen is still needed because it stimulates the egg cell to reproduce, which makes things seem as though both parents are contributing hereditary information to the offspring, but in fact only the mother's information is used. This threw off the results, and Mendel discovered that his conclusions did not hold. Not

(opposite page) Studies of patterns of inheritance for traits such as color and shape in peas revealed the principles of heredity to Gregor Mendel. Above: The "law of dominance" reveals patterns of heredity for a single gene with one dominant and one recessive allele. Below: The "law of segregation" shows that the pattern of one trait does not influence the inheritance of another.

knowing that the plant was a special case, he began to doubt his own line of thought.

By this time, Mendel had been elected abbot of the monastery. He spent the last 20 years of his life taking care of its business and performing extra duties such as taking part in city committees. This left little time for his work with plants, although he continued to work with bees. In 1869, he once again addressed the Brno Society for the Study of Natural Sciences, reporting the confusing results of his work with hawkweed and calling his own earlier hypotheses into question. After his death in 1884, it would be nearly two decades before other scientists, working much the way Mendel had, realized the true importance of his work and made his contributions clear to the world.

2

Classical Genetics (1900–1950)

Gregor Mendel had glimpsed the most fundamental principles of heredity, but when he began to doubt his own results and immersed himself in the administration of the abbey, his discoveries would be lost to the world for nearly 40 years. The rediscovery of his work would turn breeding into a science and usher in a period now called "classical genetics," marked by important discoveries about the nature and functions of genes.

As some scientists looked for patterns of heredity, others were actively searching for the physical material transmitted between parents and their offspring. Finding an answer would take another 50 years, but in the meantime, researchers working with plants and laboratory animals such as fruit flies and worms were gaining an understanding of what genes were like and how they changed through mutations. One of the great questions was whether genetics and evolution could be linked. Early students of heredity were most concerned with why species stayed the same, whereas evolutionists focused more on the question of why they changed. The difference in focus between the two fields led to a heated controversy that lasted for several decades. It would finally be put to rest with the discovery of the structure of DNA.

CELL THEORY AND THE DISCOVERY OF CHROMOSOMES

Two decades before Mendel began sorting through his first generation of peas, the German botanists Matthias Schleiden

(1804–81) and Theodor Schwann (1810–82) proved that both plants and animals were made up of cells. This discovery was possible thanks to improvements in the microscope. Joseph Jackson Lister (1786–1869), a British physicist, businessman, and optician, had changed the way lenses were made and started building instruments with two lenses. Mounting them in a tube at a precise distance from one another removed the blurriness and distortions that plagued earlier microscopes. For the first time, it was possible to see cells, and the fine structures within them, in animal tissues.

This opened the door to new way of thinking about organisms. Schleiden and Schwann did not know where cells came from, believing they might simply crystallize from the fluid in organisms. A fellow German, the physician Rudolf Virchow (1821–1902), soon proved them wrong. Looking at tissue taken from cancer patients, he suddenly realized that tumors started as single cells that divided over and over again. In another leap of insight, he saw that the same thing was true of the healthy cells that made up an organism's body. "Each cell arises from a preexisting cell," Virchow wrote in a series of lectures published in 1858.

Virchow's observations and experiments carried out by the French physician Louis Pasteur (1822–95) overturned the commonly held idea of spontaneous generation. Many scientists believed that flies and maggots grew by themselves in rotting fruit and meat—in other words, living things arose spontaneously from nonliving material. Pasteur proved that they grew from eggs. So both flies and cells had to come from preexisting organisms. This gave scientists a new starting point from which to investigate the growth of embryos and the structure of plant and animal bodies.

Another advance had been made in the late 1850s, through a combination of carelessness and good luck. It was hard to see inside cells because they did not absorb dyes well. The German anatomy professor Joseph von Gerlach (1820–96) had been trying out a number of dyes, but with little success. One evening he left a set of slides lying on a hot plate. In the morning, he was about to throw them out but decided to take a look first. Under the microscope, he saw that dyes had penetrated the tissues, giving him a clear look at cell nuclei and membranes.

Heat had been the missing ingredient, opening the door to the development of a wide range of new dyes that gave scientists their first look at new tissues and cell types. In 1879, another German biologist, Walther Flemming (1843–1905), discovered that the nuclei of cells contained threadlike structures that were split up when a cell divided. In the early 1900s, scientists learned that these chromosomes came in pairs. Each species had a characteristic number of chromosomes: Humans have 23 pairs, flies have four, dogs have

By refuting the idea that life could be "spontaneously generated," Louis Pasteur (pictured here) and the German physician Rudolf Virchow laid an important cornerstone of modern cell biology, embryology, and heredity. *(Dibner Library of the History of Science and Technology)*

39, and goldfish have 47. The differences suggested that chromosomes might contain hereditary information.

Oscar Hertwig (1849–1922), a professor of zoology at the University of Berlin, looked at the huge, pearly-white eggs of sea urchins and discovered that a sperm cell brings a new nucleus into the egg, which then fuses with the egg's own nucleus. Wilhelm Roux (1850–1924) and August Weismann (1834–1914), German university professors, figured out what this meant for heredity: Fertilization is a process of combining chromosomes from each parent. These threads, Roux wrote, must contain the hereditary material, and he proposed that the information they contained was in a linear form, like the words of a text.

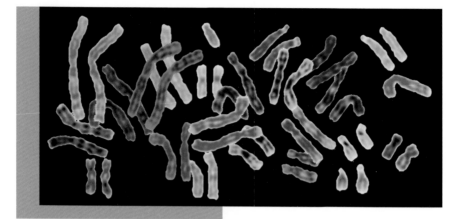

Studies of the behavior of chromosomes under the microscope in the late 19th and early 20th centuries revealed that they might carry hereditary information. Humans normally have 23 pairs, including an X-Y pair that determines whether a person will be male or female. (*National Institutes of Health*)

Weismann tried to pull all of these observations into a single theory. He believed that organisms maintained reproductive *germ cells* separate from the rest of their cells (which he called the soma), and this helped explain why organisms did not pass along traits acquired during their lifetimes to their offspring. This idea was central to evolution but was still controversial among scientists, many of whom felt that natural selection was a severe and "amoral" system. Weismann put it to the test with an experiment in which he cut off the tails of mice for several generations in a row. If Darwin were wrong, he reasoned, the mice would eventually produce offspring with no tails. But this never happened. Neither behavior nor lifetime events affected the protected germ cells.

Weismann believed the material in these cells, which he called the "germplasm," would be the key to understanding heredity. Whatever this substance was, it was passed along intact from generation to generation, separate from the rest of the body. The soma was like a flower that grew and died within a year; the germplasm was like the body of the plant, which survived season after season. The function of sex was to mix up the germplasm of separate organisms, ensuring variety within species.

THE REDISCOVERY OF MENDEL'S WORK

August Weismann proposed that each single chromosome carried a complete set of hereditary instructions. The hypothesis was wrong, but it encouraged other scientists to start their own experiments. When scientists publish their work, they need to check the scientific literature to be sure that the same thing has not been done before, and this helped in the rediscovery of Mendel's articles on heredity.

The Dutch biologist Hugo De Vries (1848–1935) believed that single characteristics, such as size or color, might be controlled by single "particles" of heredity. Working much the same way Mendel had, he began raising maize and a flower called the evening primrose. De Vries rediscovered the phenomenon that plants of the first generation all looked alike (because they had all inherited at least one dominant trait), but that in the second generation, the weaker characteristics appeared one out of every four times. This meant, he wrote, that the traits of plants were inherited in an equal number of units contributed by each parent, and that they could be dominant or recessive.

De Vries was about to publish his work when a colleague sent him a copy of Mendel's original article on peas. Later, De Vries claimed he had learned of Mendel's experiments only after having discovered the same principles on his own. He had done an immense amount of work, demonstrating that the principles held true in 20 different species of plants. Learning that his work was not original must have been a shock, but he gave Mendel credit in his paper "Concerning the Law of Segregation of Hybrids," which was published in 1900 in the journal of the French Academy of Sciences.

Two other scientists were having a similar experience. Ironically, Carl Correns (1864–1933), a German raised in Switzerland, was encouraged to become a botanist by Karl von Nägeli—the professor who had discouraged Mendel. Carrying out experiments with corn, Correns found the same type of behavior that Mendel had and came to many of the same conclusions. He wrote an article called "G. Mendel's Law Concerning the Behavior of the Progeny of Racial Hybrids," which appeared two months before De Vries's work.

The third "rediscoverer" was an Austrian, Erich Tschermak von Seysenegg (1871–1962), who also had a biographical connection to Mendel. Tschermak's grandfather had been one of Mendel's teachers at the University of Vienna. The grandson had read of experiments carried out on peas by Charles Darwin. He began to work with the plant, also repeating Mendel's work without realizing it. He came to the same results and wrote them up in his university dissertation, which was published in the Austrian journal of experimental agriculture. But Tschermak did not immediately understand the significance of what he had discovered. He did not realize, for example, that the 3:1 ratio meant that genes had dominant and recessive characteristics; he described some of the traits as having more "hereditary potency" than others. Tschermak's aim was to develop better crops, and eventually he did, creating strains of wheat, barley, and oats that were more productive than the existing ones.

These three rediscoveries clearly demonstrated that Mendel's principles held true in a wide range of plants beyond peas. Even so, wide recognition for Mendel and his work among the scientific community was truly achieved only by biologist William Bateson (1861–1926), who was carrying out his own crossbreeding experiments in England. While riding on a train to give a lecture in London, Bateson read De Vries's paper with its reference to Mendel; he immediately realized that Mendel's laws laid the groundwork for an entirely new science of heredity. He also believed that they would provide a solid foundation for evolution by showing why variety existed within species.

Bateson became a passionate advocate for the new science. He had Mendel's original paper translated into English so that it would be widely read. In 1902, he wrote a book called *Mendel's Principles of Heredity: A Defence* and made sure that scientists across the globe got a copy. The sudden flood of evidence for Mendel's ideas revolutionized the way researchers thought about heredity. Within 10 years, scientists were investigating the laws in a wide range of plants and animals and had started to put them to practical use. In the United States, large agricultural stations were set up to create better corn, beans, wheat, tobacco, and other crops based on genetic principles.

Bateson and his students began a wide variety of experiments that would refine and expand on Mendel's work. His work with chickens proved that Mendel's rules held true for animals as well as plants. Another important discovery was that not all traits followed the 3:1 ratio. Colors in flower petals, for example, were passed along at a ratio of 15:1. Mendel had explained the meaning of the number 16 when he combined different traits (color and shape in peas), and Bateson understood its significance for flowers: Making color in a petal required two traits, rather than one. Not everything about an organism was the result of a single hereditary unit.

Bateson began to invent a new language to fit the new science, which he called genetics. In an organism, each trait was present in two copies that he named allelorphs (later shortened to *alleles*). If the alleles inherited from two parents were identical (for example, if both were dominant, or both were recessive), the pair was called *homozygous*. If the pair consisted of a dominant and recessive trait, it was termed *heterozygous*. The word *gene* was first used a few years later by the Danish botanist Wilhelm Johannsen (1857–1927) to refer to the units.

Johannsen also invented the terms *genotype* and *phenotype*. The phenotype referred to all the physical characteristics present in a specific organism—from large, visible traits such as the color of its eyes or its behavior, all the way down to the chemical makeup of cells. The genotype referred to an organism's total hereditary material. The two were often different because organisms often had genes for traits that never appeared in their bodies. A round pea might carry a recessive allele for

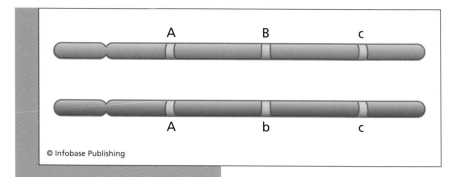

© Infobase Publishing

Animals inherit two copies of each chromosome, which usually contain alleles of the same genes. The images above show the two copies of one of an animal's chromosomes. A, B, and C are genes; capital letters represent dominant alleles, and small letters represent recessive alleles. The animal is homozygous for gene A because it has two copies of the dominant allele. It is heterozygous for gene B because the allele on one chromosome is dominant, and on the other it is recessive. Because the animal has two recessive copies of c, it is homozygous for gene c.

wrinkledness. Women passed along genes for baldness to their sons without becoming bald themselves. A particular phenotype might disappear for a generation, or even skip several generations, while remaining present in the genotype.

Evolution posed a problem for this first generation of geneticists. Mendel's rules showed how alleles were passed down through families, but shuffling around genes could not explain the great changes that species had experienced during evolution. Peas could be round or wrinkled; Mendel's experiments did not say anything about how a third kind—say, a square pea—could arise. One explanation might be that an unusual combination of recessive genes was required to create a feature. But this did not go far enough. Humans surely had more genes than simpler animals, or plants, or the single-celled organisms they had evolved from. Where had the extra information come from?

The question would not really be answered until scientists learned much more about the nature of genes. But in 1901, De Vries came up with a hypothesis that proposed part of the

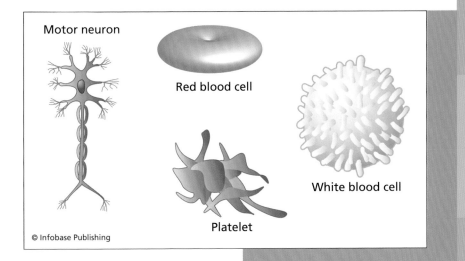

Motor neuron

Red blood cell

White blood cell

Platelet

© Infobase Publishing

solution: Sometimes there are sudden, "discontinuous" changes in one of the alleles of a gene—an event that does not obey the normal rules of heredity, like a spelling mistake that might creep in as someone copies a text by hand. De Vries called this process mutation.

A genotype is the complete collection of alleles in an organism's genome. The phenotype comprises the features it exhibits. A few examples can illustrate the difference: (A) Cells taken from the same person have the same genotype but many different phenotypes, such as the motor neuron, red blood cell, platelet, and white blood cell shown above. (B) Identical twins have the same genotype but slightly different phenotypes (e.g., their fingerprints are different). (C) A woman may carry a recessive gene for color blindness and pass it along to her sons without being color blind herself.

CHROMOSOMES AND HEREDITY

As Mendel's ideas gained acceptance, cell biologists were following up on Weismann's chromosome theory, hoping to discover the physical substance that genes were made of. In Munich, Theodor Boveri (1862–1915) pursued the question in a series of experiments with the eggs of sea urchins, which were large, transparent, and easy to study under the microscope. Genes had to be in the nucleus, Boveri figured, because they were delivered to the egg by sperm, and sperm were little more than a nucleus with a tail attached. He

The Debate over Natural Selection

The discoveries of the early 20th century triggered a debate as to whether Mendel's genetics could be compatible with Darwin's theory of evolution. The biggest disagreements concerned where variety in species came from and the importance of natural selection. Some geneticists, including William Bateson, thought that shuffling a population's genes and introducing a mutation from time to time might be enough to make a species "drift away" from its original form. This might lead to evolution without any assistance from natural selection. Bateson collected hundreds of examples of what he called *discontinuities:* cases where plants and animals undergo strange duplications of their parts. He found insects with extra body segments, giving them extra pairs of legs, and goats with extra horns. Over time, these changes might work their way through the population until it became a new species, without any help from natural selection. This idea appealed to many scientists and the public, who felt that natural selection made nature an unnecessarily violent place.

But natural selection was such a key part of the theory that Darwin's strongest supporters rejected Bateson's idea. His monstrosities, they said, would quickly be eliminated in favor of well-adapted members of the species. They believed that evolution worked on more subtle, measurable characteristics of organisms, such as size, the size of animal litters, and the numbers of seeds produced by plants—things that obviously could give an organism an advantage at survival and reproduction. They created the field of *biometrics:* measuring anything about an organism that could be measured, plotting characteristics such as size, weight, and strength on charts, and studying curves of distribution. Evolution would be seen, they claimed, as the middle of various curves gradually shifted from one generation to the next.

The first chapter described how Darwin thought of heredity as a process of blurring rather than mixing and matching discrete characteristics. This idea persisted into the 20th century; evolutionists did not yet see how genetics could explain both the small differences within species and the large differences between them. Many geneticists had the same problem. Today this is understandable because both the most subtle and the most dramatic mutations that occur in a species are invisible. Most changes in the genetic code have small effects on cell chemistry, which were impossible to analyze until the late 20th century. Others are so dramatic that an organism does not live long enough to be studied. Geneticists could study only visible changes in plants and animals that survived.

It took a group of "outsiders"—mathematicians—to bring the two points of view together in the first decades of the 20th century. Three Englishmen—Ronald Fisher, Sewall Wright, and J. B. S. Haldane—used mathematics and models to prove that natural selection was a much more powerful force for species change than "drift." Even if a mutation gave an organism only a slight advantage in reproduction over other members of its species, a new trait would spread quickly through a population. Still, the debate really ended only in the 1950s, when chemists, physicists, and biologists created a model of DNA that could explain both normal heredity and the reasons for mutations. A single change in the genetic code could push an organism to one side of the "curve" of species variety, could create Bateson's monstrosities, or might have subtle effects that could be detected only by studying the chemistry of cells. Any of these changes would be subject to natural selection.

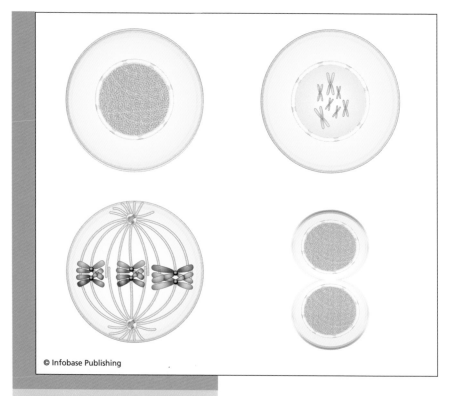

In the resting cell (upper left), DNA is sprawled loosely through the nucleus like a tangled thread. After it is copied and the cell prepares to reproduce, it condenses into huge chromosomes (upper right). These line up in the center of the cell, where they can be grabbed by the mitotic spindle (lower left). They are separated equally to give each of the two daughter cells a complete set of the organism's DNA (lower right). This behavior convinced many scientists that genes were located on chromosomes.

showed that if more than one sperm managed to fertilize an egg, the resulting embryo had too many chromosomes, failed to develop, and died at a very early stage. Fertilized eggs with the normal set of chromosomes produced healthy embryos.

This meant that there was more to heredity than just possessing all the necessary chromosomes; otherwise, why would it hurt to have extra ones? Boveri realized that an organism needed a particular combination to develop in the right way. Experiments conducted between 1901 and 1905 made him realize that each chromosome possessed unique qualities. Weismann had been wrong: Each chromosome contained a different part of the instructions needed to build an organism. Too many chromosomes meant too many instructions, and too few meant that important information was missing.

In the 1890s and early 1900s, American scientists began to make important contributions to cell biology and genetics thanks to the creation of new science departments in several American universities. They were set up to host professors' laboratories and were excellent places to combine education and research. A good example was Bryn Mawr, a prestigious college for women in Pennsylvania, where Edmund Beecher Wilson (1856–1939) was hired to start a department of biology in 1885. Wilson moved to Columbia University six years later, and the network of students and colleagues that arose around him at these two universities over the next decades had a huge impact on the development of genetics worldwide.

Wilson began his career studying how embryos develop but changed his focus after visits to Europe, where he became close friends with Theodor Boveri. He followed up on some of Boveri's experiments, taking advantage of new types of microscopes and staining techniques. In 1896, this work led him to propose the radical new idea that a molecule called nucleic acid—DNA—carried the hereditary material. Although chemists had isolated DNA from cells almost 30 years earlier, its functions were not understood.

Wilson's idea was not widely accepted for two main reasons. First, when chromosomes were broken down with chemical methods, they contained a lot more than DNA. Nucleic acids were entangled with thousands of types of proteins in a very complex mixture called *chromatin*. Second, the chemical recipe of DNA was too simple, most scientists thought, to carry "complex" information. Proteins seemed better candidates—they were built of a much larger chemical "alphabet" that was surely sophisticated enough to carry complex information. Wilson's intuition would take more than 50 years to prove.

SEX AND THE X-Y CHROMOSOMES

Walter Sutton (1877–1916) joined Wilson's group at Columbia as a graduate student in 1902. His main interest was to discover how heredity produced organisms of two sexes. He had not yet heard of Mendel's laws, but his own observations of cells hinted that parents contributed equally to the characteristics of their offspring and that hereditary material consisted of units on the chromosomes.

As an undergraduate at the University of Kansas, Sutton worked on chromosomes in the laboratory of Clarence Mc-Clung (1870–1946), who believed that an extra chromosome made animals into males. At the time, little was known about chromosomes except that they appeared suddenly at the onset of cell division and then vanished again. The reason behind this mysterious phenomenon lay in the fact that DNA is normally a huge, loosely-strung tangle in the cell nucleus, a thread that is too thin to see even with the most powerful light microscopes. (The strand can barely be detected even with an electron microscope, which was invented decades later.) Shortly before cell division, the strands condense by pulling together in thick clumps that can be clearly seen when stained.

By carefully studying the chromosomes of grasshoppers and other insects, Sutton discovered that chromosomes always took on the same shapes when they re-formed. This provided a way to tell them apart and track them through phases of the cell lifecycle. It supported Boveri's idea that each chromosome contained a different part of an organism's total genetic material. The two men went further and proposed that chromosomes were subdivided into additional units that carried specific traits—the way a sentence is a long string of words. If so, it might be possible to match traits to chromosomes. One of them might carry the information that determined an animal's sex.

Sutton was investigating a species of grasshopper that had 22 chromosomes—11 pairs—but in half of the sperm he found a 23rd "accessory," or "X," chromosome. Its function was discovered by one of the first women to receive a Ph.D. in science in the United States, Nettie Stevens (1861–1912). She received her Ph.D. at Bryn Mawr in the laboratory of the famous American geneticist Thomas Morgan (1866–1945), then spent a year abroad in Boveri's laboratory. Upon returning she turned her attention to a species called the mealworm and the same problems Sutton was working on. She discovered that females had 20 large chromosomes, whereas males had 19 large chromosomes and one smaller one. When she looked at their sperm, she discovered that those with 10 pairs of large chromosomes produced females; those with nine and a small chromosome ("Y") in the 10th pair became males. At about the same time, E. B. Wilson was finding the same phenomenon in the chromosomes of several species of insects. They had found the reason for the existence of two sexes in humans and many other species.

A newborn animal needs a full set of chromosomes. These are provided by an egg and a sperm, so these two cells possess a half set, but they are originally made from immature reproductive cells with pairs of chromosomes that are split up. Since a male has an X-Y pair, half of the sperm his body produces receive a single X chromosome and the other half a Y. Eggs are made by females, who have a pair with two X chromosomes, so each egg receives an X. The sperm determines whether a baby is a boy or a girl—so it was Henry VIII's fault that he kept fathering girls, and not that of his many wives.

The genetics of sex in some other species works differently. In Sutton's grasshoppers and some other insects, males have only a single X chromosome and lack the Y entirely. Male birds have one pair with identical chromosomes (ZZ), whereas females have a nonidentical pair (called ZW). Male and female alligators have identical chromosomes; sex is determined by environmental factors. Whether an embryo becomes male or female depends on the temperature at which the egg is incubated. And in 2004, scientists finally solved one of the oddest cases of sex determination—that of the platypus. This strange, egg-laying mammal has 26 pairs of chromosomes. Males have five XY pairs, which are XX in females.

FRUIT FLIES AND THE BIRTH OF THE MODERN LABORATORY

Upon his move to Columbia in 1891, E. B. Wilson picked a talented young zoologist named Thomas Morgan to replace him at Bryn Mawr. Throughout his early career, Morgan had a wide variety of scientific interests. He had worked on cell biology and carried out experiments on pigeons, rats, and lice. After finishing his Ph.D., he spent a period abroad in Hugo De Vries's laboratory in Amsterdam, where he became interested in the question of mutations. After several years at Bryn Mawr, Morgan followed Wilson to Columbia.

Morgan was skeptical about nearly everything until he had seen it with his own eyes. He questioned Darwin's theory of natural selection as the main cause of evolutionary change, whether Mendel's laws of heredity worked in animals, whether genes were really located on chromosomes, and whether the X-Y chromosome pair determined sex. He often repeated other scientists' experiments before becoming convinced himself. He ended up carrying out research projects in embryology, cell biology, and evolution, often in collaboration with Franz Lutz at Harvard University.

Visits to De Vries's laboratory had convinced Morgan of the Dutchman's idea that species remain stable for a long time, then suddenly undergo spurts of mutations that create new species. If this were true, it ought to be possible to catch evolution "in the act" in the lab, but one would need immense patience and the right experimental animal. Nettie Stevens, William Castle, and Lutz had been working on the fruit fly, an annoying insect that fed on yeast molds that grew around trash containers, fruit, or a glass of wine left in the open. *Drosophila* had only four chromosomes, which might make it easier to link them to heredity. Following a piece of advice from Lutz, Morgan created a fly lab.

This decision was an incredibly important event in the history of genetics. Most work was being carried out with plants because they were cheaper and easier to work with than animals. There were no cages to clean, and they produced far more offspring than mice or guinea pigs. On the other hand, plants obeyed the growing seasons, so work was often limited to one generation per year. A fruit fly could reproduce just two weeks after birth and was easy to manage in the lab. Thousands could be kept alive in a glass jar, fed on mashed bananas. Flies did not require the delicate care and handling of plants, so students could be easily trained to work with them. Many of the pioneers of *Drosophila* genetics were students, including a number of women, who could work as technicians or assistants but had very few opportunities to obtain advanced degrees in the early 20th century. And the work with flies created a team model of doing research that has continued up to the present day.

Decades of work on the fruit fly by Thomas Hunt Morgan and his colleagues led to the discovery of hundreds of genes and revealed crucial aspects of their structure and functions. Here, Morgan is shown at work in his laboratory at the California Institute of Technology. *(A. F. Huettner/the Archives, California Institute of Technology)*

One of the first experiments carried out in Morgan's lab was a test of Lamarck's ideas about heredity. Graduate student Fernandus Payne (1881–1977), who had recently arrived from Indiana University, had been working with a species of blind fish found in caves that had no light. While Lamarck's theory held that living in darkness was itself the cause of blindness, evolutionists had another solution: Eyes were sensitive, exposed organs that could be wounded and infected. In a completely dark environment, eyes would have no beneficial value, but if chance produced a fish without them, it might have better chances of survival than its counterparts. Fruit flies could be used to test the competing hypotheses. Payne began raising them in the dark, but even after doing so for 69 generations in a row (described in a paper called "69 Generations in the Dark"), he did not obtain any blind *Drosophila*. He gave his flies back to Morgan, who was becoming frustrated as he waited for unusual bursts of mutations to happen in chickens and other animals.

In the meantime, Lutz was making progress in studying the insects. In 1907 and 1908, he discovered mutations that changed the structure of the veins that run through the fly's wings; another produced dwarf-sized flies. Things were going slower for Morgan: It took two years for him to begin to find mutations. Still hoping to prove De Vries's idea that species evolved because of changes in the environment, he had been trying to cause mutations by exposing insects to radiation, changing their diets, and other methods, but for a long time none appeared. That seems unusual today because mutations are frequently found in fruit flies even under normal circumstances, but Morgan may have needed to gain experience in looking at the insects before he was able to recognize subtle changes. And originally the laboratory may not have had enough flies—mutations become likely in populations of a few thousand; Morgan was working with hundreds. The process was frustrating, and he may have been about to give up. Suddenly, in January 1910, he noticed a subtle change in color patterns on the insects' bodies. At that point the floodgates opened, and mutations began appearing everywhere.

The first really dramatic find was a male fly whose eyes were white rather than the normal red color. If it was a mutation, the trait ought to be hereditary, so Morgan bred the fly with others and obtained both males and females with white eyes. The next step, crossing these with other animals, produced results that at first seemed confusing because they did not follow the normal pattern of Mendelian inheritance. When he crossed red-eyed males with white-eyed females, the sons all had white eyes and the daughters red. On the other hand, white-eyed fathers and red-eyed mothers produced a first generation that all had red eyes. When these offspring mated, the second-generation *male* flies showed a 3:1 ratio of red to white. Mendel's rules were working, but the sex of the fly was somehow skewing the pattern.

Morgan's colleague E. B. Wilson solved the mystery by proposing that the gene for eye color might be located on the X chromosome. Females had two copies of any gene located there (because they had two X chromosomes), and males only one—inherited from their mothers. Somehow, a single copy of the mutant gene gave males a trait that did not appear in females. Wilson suddenly realized that the same phenomenon might explain something he had observed in human inheritance. He was extremely color blind and had been studying how this was inherited in families. His genealogies suggested that this trait, too, was passed from mothers to sons. The same turned out to be true of hemophilia and many other genetic diseases. The responsible genes might be located on the X chromosome.

A month later Morgan discovered a second mutation linked to sex: wings that grew to only about half the normal length. Then, a talented artist and researcher in the lab named Elizabeth Wallace, who was making painstaking illustrations of the flies, found a mutation that gave males yellow bodies. Another created vermillion-colored eyes. Some of these features were related to sex; others were not.

Morgan began a tradition of naming genes after the effects of mutations. For example, he called the gene that caused white eyes "white," although the function of the normal version of the gene was probably to create red eyes. Mutations that changed the size and shape of wings were called "miniature" and "truncate." By about 1915, the lab was working with dozens of different strains. Crossing them with each other, the scientists began to unravel much more complex patterns of inheritance.

As more mutations appeared, the focus of work in his lab moved from trying to catch evolution in the act to focusing entirely on heredity. Morgan began to think in a different way about genes and their roles in the life of an organism. Previously, a main goal of genetic studies had been to discover how many genes were needed to build particular organs and larger structures in animals, such as the eye. Understanding that, scientists hoped, would reveal how small changes in such genes could lead to the evolution of structures and species. The work of Morgan's lab suggested that such questions might be far too complex to answer without first gaining a far better understanding of genes. Even that was a huge project: No one knew how many genes a fly had, or how they worked together.

GENE MAPS

Mendel showed that, in principle, genes were inherited independently: The color of a pea had no influence on whether it was round or wrinkled. But as Morgan crossbred increasing numbers of mutant strains, he began to discover exceptions. The reason had to do with the physical locations of the genes. It was true that genes on different chromosomes followed independent patterns of inheritance. But genes on the same chromosome were usually inherited together. This was easiest to determine in the case of sex-linked genes, such as those that produced white and yellow eyes—which were located on the X chromosome.

Since a male fly got its one X chromosome from its mother, it inherited every gene on the X chromosome from her. Collections of genes on other chromosomes were also inherited as a package. Working with dozens of strains and crossbreeding tens of thousands of flies, Morgan's lab had been collecting a huge amount of data about patterns. Curiously, the "rule" did not always work. In rare cases, genes on the same chromosomes were not inherited together. In 1912, Alfred Sturtevant (1891–1970), one of Morgan's students, proposed an explanation.

Sturtevant's idea had to do with the way sperm and egg cells were created. Unlike most cells, which are created with pairs of chromosomes, reproductive cells have only one chromosome from each pair. They are made in a special form of cell division called *meiosis.* Before pairs are split, they line up side by side. When a Belgian researcher named Frans Alfons Janssens (1863–1924) watched this happen under the microscope, he saw that the strands twisted around each other, making sharp bends. Morgan thought that the pressure would make the two strands break at the same sites. Cells had to have some way to repair them, but in the process, pieces of the neighbors might be exchanged, and genes might be transferred to another chromosome. Morgan called this process crossing over; today it is known as *recombination*. It is an important source of the variety within species. Mothers do not pass down entire, intact sets of chromosomes to their daughters, who then transmit the same chromosomes to their own daughters. The information in a chromosome is remixed with each generation.

Until now, genes had remained abstract ideas; Morgan felt that it made little difference what genes were made of, as long as they behaved properly. That was about to change. In a discussion with Morgan, it suddenly occurred to Sturtevant that recombination could help scientists pinpoint the positions on chromosomes.

His idea went something like this: Suppose that genes were like words that appeared in a few very long sentences (chromosomes). Each mutation discovered by the lab represented a word. Making a gene map would be like trying to reconstruct the sentences. Fortunately, the words did not come completely separately and randomly, but in blocks—genes that were inherited together. Crossbreeding mutant strains showed which words "belonged to the same sentence," and this would allow scientists to assemble groups of genes into a map.

To take the metaphor further, suppose that the sentences were printed over and over on long strips of paper. Recombination was like cutting each strip at a random place. Just as words close to each other on a strip would often stay on the same half when it was cut, neighboring genes were more likely to be inherited together. Words far apart from each other in the sentence were more likely to be separated when the paper was cut.

Sturtevant collected Morgan's data that showed the frequency at which genes were inherited together, took it home, and in a single night managed to plot six genes. The map showed their order on the chromosome and gave a relative idea of their distances from one another.

It was easiest to map the X chromosome, but the group quickly moved on to others. By 1915, they had plotted 36 genes on four chromosomes. By 1926, they had found 36 genes on one chromosome alone. In the process, they had changed the nature of genetics. This young science was no longer simply a sophisticated analysis of breeding, nor did it aim to show how many genes were required to build complex organs such as eyes. Instead, its focus had become the structure and behavior

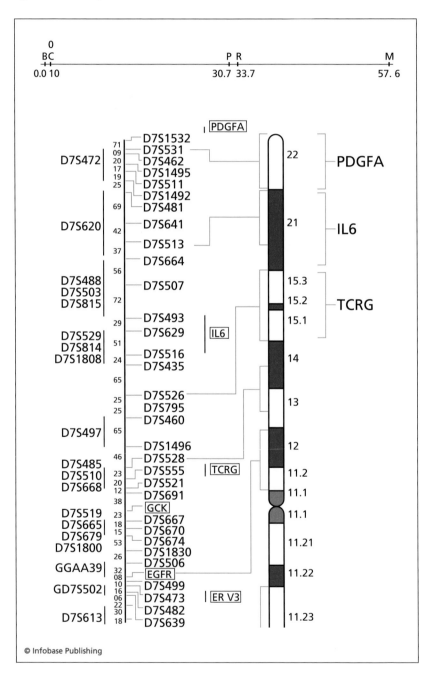

of genes, discovering new ones, mapping them onto chromosomes, and understanding how they worked together in chromosomes to transmit the hereditary material.

For a few years, Morgan's laboratory at Columbia University was a sort of Camelot of the new genetics, but other universities were catching on and hired some of his students as professors. Rice University in Texas took in Hermann Muller, who established a new fly center there. Cornell University launched a fly lab, in addition to a program dedicated to the genetics of maize. Cold Spring Harbor Laboratory in New York created another center for fly genetics. Then, in 1928, Morgan and a large part of the group moved to the California Institute of Technology.

CHROMOSOME PUZZLES

Calvin Bridges (1889–1938) began his career washing bottles and preparing food for the flies in Morgan's lab at Columbia. He was clever at laboratory work, inventing techniques such as using ether to anesthetize the flies and watercolor brushes to sort them on porcelain plates. When Bridges discovered an interesting mutant with brightly colored eyes, Morgan gave him a desk and brought him on to the scientific team. It was a wise move: Bridges quickly proved that he was as clever at intellectual puzzles as technical ones. One of his first discoveries showed that in very rare cases, traits located on the X chromosome were not inherited in the expected way. White-eyed female flies mated to red-eyed males should produce only white-eyed sons, but about one in 1,000 of the male offspring had red eyes. He called this *nondisjunction* and interpreted it correctly: Sometimes egg or sperm cells received two copies of the sex chromosome, or none, rather than one. Later, similar cases would be discovered in humans.

(opposite page) Above: The first gene map, made by Sturtevant in 1911, showing the relative positions of six genes on the X chromosome of the fruit fly. Below: A small section of a modern gene map, revealing a small proportion of what is known about the genes on part of human chromosome 7.

Morgan and his group were working with such huge numbers of flies that nondisjunctions and other rare exceptions to the rules of heredity kept turning up. Bridges had a talent for figuring out what they meant. First, he discovered that subsections of chromosomes sometimes disappeared—for example, a female might be missing a piece of one of its X chromosomes. The group's gene maps made it possible to identify the positions, and sometimes the sizes, of these missing pieces—which could range from a piece of a single gene, to blocks containing many genes, to entire chromosomes. They could be detected because such losses broke the rules of Mendelian inheritance for the genes involved.

Then Bridges found that pieces of chromosomes containing one or more genes were sometimes duplicated. This was an extremely important finding because it showed where extra genetic material might come from over the course of evolution—humans and other complex forms of life had many more genes than their one-celled ancestors. Today, scientists know that once such duplications occur, genes and their copies undergo different mutations, which often leads to the development of different functions. This supports the idea that the genetic material found in all of today's organisms started off as a small set of genes that underwent duplication after duplication and countless mutations over billions of years.

Progress in finding and mapping genes was limited by the number of mutations the scientists had to work with. By the 1930s, Morgan and his group had discovered several hundred genes in the fly, but this was probably only a small percentage of the real number. Morgan once guessed that the fly might have 2,000 (he underestimated; the completion of the *Drosophila* genome revealed about 14,000). Scientists had to wait for mutations to happen; they could not cause them. That changed when Hermann Muller, at his new laboratory in Texas, began trying out methods to cause more mutations in the flies. Radiation greatly increased their frequency—a discovery for which he received the Nobel Prize for physiology or medicine in 1946.

The laboratories of Muller, Morgan, and others exchanged information and strains of flies, but at the same time, groups competed fiercely to be the first to make key discoveries. Competition was particularly strong between the labs of Muller and Morgan, possibly because Muller resented how he had been treated at Columbia—he had never been completely accepted as a member of the group. Tension reached a high point in the 1930s, when Bridges (who had moved to Caltech with Morgan) made a huge leap forward thanks to the work of Theophilus Painter (1889–1969), a member of Muller's laboratory.

Painter had discovered that cells in the fly's salivary gland contained giant-sized versions of chromosomes. They were so large that they could be observed in great detail under the microscope, and he began staining them with dyes. This created dark bands on each chromosome that appeared in the same places every time it was stained. This was of huge interest to gene mappers. While they knew which chromosome each gene was located on and its relative distance from other genes, its position could not really be seen.

The task was like trying to draw a map of the United States using only a table of driving distances between cities. The result might be very precise, but it still might be impossible to overlay the map on a satellite photograph. Similarly, the exact physical locations of genes on the chromosomes were not known. Chromosome bands might solve the problem.

This required a landmark, and Bridges knew of a gene that might provide the key. Years before, Morgan and Sturtevant had discovered a gene they named Bar because mutations gave the flies thin, rectangular eyes. In some flies, the effect was mild; in others, it was very strong. Morgan and Sturtevant thought that the most dramatic effects happened in cases where the mutant gene was duplicated, giving flies more than one copy. It was located on the X chromosome, and Bridges had calculated its relative position.

Now he stained giant chromosomes from flies with the Bar mutation and studied them carefully under the microscope. The pattern of bands was slightly different in normal and mutant insects—demonstrating the location of the gene. Just as he expected, in flies with the narrowest, slotlike eyes, this region was duplicated. Muller, during a stay in Russia, was carrying out exactly the same experiment, with the same results, and a dispute erupted over who deserved credit. Both sides felt strongly about the issue because of the importance of the discovery: In a few experiments, the men had proved that a gene's physical position could be pinpointed, that duplications of genes occurred, and that having two copies of the same gene had an impact on how the eye formed.

Gene maps meant that heredity could now be studied in cells under the microscope as well as in living creatures. Unfortunately, the method used to stain chromosomes in flies did not work in the cells of mammals. It was not until the 1970s that geneticist Torbjörn Caspersson (1910–97) and his colleagues in Sweden developed a technique to reveal bands in human chromosomes. This permitted the first chromosome maps to be made showing the positions of human genes and resulted in another piece of evidence for evolution: In many cases, genes appeared in the same order on human and fly chromosomes.

MAIZE AND "JUMPING GENES"

At Cornell University, a young woman named Barbara McClintock was trying to accomplish the same thing with the chromosomes of maize. She was an excellent microscopist who had proved that the plant had 10 chromosomes, and at Cornell she assembled a team of plant breeders and cell biologists to make maps of maize genes. Rather than staining, she relied on the fact that the plant's chromosomes had knoblike bumps that appeared in regular places on the chromosomes. They, too, could serve as landmarks.

McClintock's sharp eyes and her deep familiarity with the features of maize allowed her to make huge leaps forward in understanding what genes were and how they worked. She was so far ahead of her time that it took several decades for many of her discoveries to be accepted by the scientific community. Her contributions were finally recognized with a Nobel Prize in physiology or medicine in 1983. The prize in this category is usually awarded to two or three individuals in a single year; so far, McClintock remains the only woman to win it alone.

McClintock's methods clearly showed what happened to specific genes during reproduction. She drew pictures of the strange shapes that chromosomes formed as they intertwined, broke, and exchanged pieces with each other (the "crossovers" predicted by Morgan). The twisting sometimes made regions of neighboring strands run in opposite directions. When these regions broke and were recombined, a fragment might be pasted into the neighboring chromosome backward. As Calvin Bridges hypothesized that genes might undergo such *inversions* based on studies of mutants, McClintock was watching them happen under the microscope.

Moving on to the University of Missouri in Columbia, McClintock began using X-rays to cause mutations in the plant. Radiation had a curious effect on maize chromosomes. They were normally shaped like small sticks, but when damaged, their ends sometimes joined together and formed rings. McClintock deduced there had to be structures on the ends of chromosomes that normally prevented this; radiation damage kept them from doing their jobs.

Barbara McClintock's pioneering studies of heredity in maize revealed aspects of gene behavior that would not be understood or accepted for nearly 50 years. Here she is shown at Cold Spring Harbor Laboratory in the 1930s. *(Barbara McClintock Papers, American Philosophical Society)*

She called the structures telomeres, and today they are known to play an important role in aging, cancer, and some genetic diseases.

McClintock also made some surprising discoveries about genes themselves. For example, by studying patterns of colored kernels in corn, she claimed that during reproduction, genes sometimes jumped from one position to another on the chromosome. She called them *jumping genes,* or *transposons.* She also discovered that particular regions of chromosomes could have a "controlling" effect on others—in other words, they could influence whether a gene was actively used by cells or not.

Most of these ideas were a radical departure from the way other scientists believed that genes and chromosomes behaved. A great deal of work would have to be done before McClintock's peers became convinced. First, they had to understand how information in genes was used in living beings. Scientists knew that the genetic code was more than a library of books, passed down from generation to generation. It was also a set of instructions for building cells, tissues, and whole organisms.

"ONE GENE MAKES ONE ENZYME"

A young Nebraskan farmer named George Beadle (1903–89) came to study at Cornell during the period that Barbara McClintock was making the first gene maps of maize. McClintock taught Beadle techniques that helped him complete a Ph.D. in maize genetics; then, like so many other scientists of his generation, he went off to work in Morgan's lab. But he was dissatisfied at Caltech because Morgan had little personal interest in the chemistry of genes. For Beadle, the subject was crucial, because he felt that ultimately life had to be governed by physical and chemical processes, and the goal of biology ought to be to explain them. The substance of genes had to determine their function: to tell cells how to develop and form a properly built animal. Despite this philosophical difference, Morgan recognized the importance of Beadle's work and gave the young man's career an important boost.

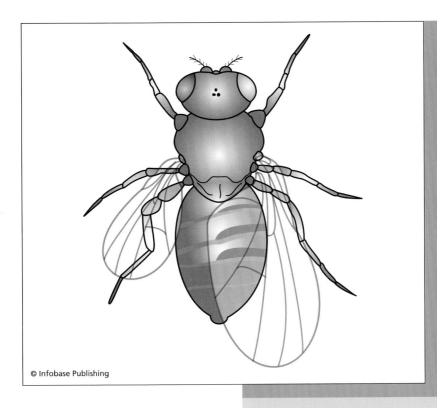

© Infobase Publishing

And many of the projects going on in the lab were providing useful hints about the nature of genes. Alfred Sturtevant had just discovered flies called *gynandromorphs* that

Among the strangest mutants found by Thomas Morgan and his colleagues were gynandromorphs—flies that were half male, half female.

grew in a very odd way: Half their bodies became male, and the other half female. The reason was complex and unusual. Gynandromorph embryos began as normal female eggs, with two copies of the X chromosome, but very early on some cells lost one copy, and other cells lost the other. As they continued to divide and form the growing body, their descendants inherited whichever copy remained. Because the two chromosomes often contained different versions of a gene (different alleles), this led to bodies that were partly controlled by one set of genes, partly by the other.

Morgan's lab had proven that genes for eye color were located on the X chromosome, which meant that a single gynandromorph might have left and right eyes with different colors. But that never happened, and Sturtevant thought he knew why. Even if a cell had the mutant gene for white eyes, actually building eyes of that color might require additional instructions from genes in neighboring cells. In some cases, those instructions, passed along in the form of small molecules, might even override a cell's own genetic information. How much of an animal's development was steered by such molecular "conversations" between cells, and how much was the result of a cell's own genes? Curt Stern (1902–81), a German geneticist who joined the lab after moving to the United States before the outbreak of World War II, had developed a technique that might give an answer.

The very first cells in a growing embryo are identical, but very soon they begin to specialize. They develop into clusters of cells called *imaginal disks;* one disk develops into an eye, another into a wing. Each part of the insect begins this way. Stern found that these bits of tissues could be removed and transplanted between flies. Watching how an imaginal disk developed ought to reveal whether it carried all the information it needed to build the structure, or whether it needed extra instructions from surrounding tissues. For example, an imaginal disk from a mutant fly could be transplanted into a normal insect, then one just had to wait and see whether the mutant or the normal trait emerged.

In 1934 George Beadle and Boris Ephrussi (1901–79), another short-term visitor to Morgan's lab, decided to tackle the problem using Stern's method. When Ephrussi moved to Paris, Beadle joined him there. They spent months peering into the dual lenses of a binocular microscope—an instrument with two sets of eyepieces focused on the same sample, a bit like a car used in driver education. They had to carefully remove bits of tissue from one embryo and transplant them into another. It was tiring, challenging work that required enormous patience and four hands, one preparing and holding the larvae, the other removing tissue and implanting it. In very early embryos, it was nearly impossible to tell the difference between imaginal disks for different parts of the fly, so the two men had to wait for the flies to grow to see what they had transplanted.

A breakthrough came when one summe[
found a third eye growing in the middle of the[
of the transplants. It meant they were on the[
had found and moved an imaginal disk for the [
months, Beadle and Ephrussi got so good at [
they could carry out 200 of the microscopic operations p[

The results were fascinating. Putting an imaginal disk for the eye from a mutant into a normal embryo always led to a fly with normal, red eyes. This meant that the tissue surrounding the transplant was able to "rescue" the defect—in other words, information for color was not contained in the eye disk at all; it was coming from outside. This suggested the next experiment: What would happen if they put a mutant disk into another type of mutant? They did this for flies with vermillion- and cinnabar-colored eyes. The results varied depending on which type of disk they put into which type of fly.

Soon it became clear that several genes worked together, in a particular order, to create the normal red pigment. Beadle and Ephrussi learned the order by studying which piece of information could override the other. The gene for cinnabar, for example, acted after the vermillion gene, because vermillion information in the imaginal disk was overwritten when it was transplanted into a cinnabar mutant. And a cinnabar disk in a vermillion fly also became cinnabar. This confirmed what scientists around the world were coming to believe: The function of genes was to operate the chemistry of the cell, and multiple genes often worked together in a series of steps, a pathway, to accomplish one thing.

These facts brought Beadle closer to his goal of discovering the chemical nature of genes. The major types of molecules in cells were known: proteins, DNA, a similar molecule called *RNA,* and *lipids* (the fats that make up membranes). Proteins carried out most day-to-day tasks, chopping up other molecules, pasting them together, or triggering various types of chemical reactions. They had such important roles that many researchers believed that genes were made of proteins.

Yet Beadle remained unconvinced and wanted a much deeper look at genes' roles in cell chemistry. To do so, he needed a pler animal—it took a long time to understand the chemistry of mutations in a complex animal. And many mutations—possibly most—remained invisible, because they killed embryos at a very early stage, leaving no mutant fly to study. When Beadle took a new position at Harvard, he hired a chemist named Edward Tatum (1909–75) to help him work with a type of bread mold called *Neurospora*. They designed a brilliant experiment to look for new genes and find out how they worked.

The first thing they did was to expose the mold to radiation, causing mutations. Then they grew it in cell cultures that had only low amounts of the substances that it needed to survive. A cell with normal genes and normal chemistry could manage to scrape up enough nutrients to survive, but mutants might not. Beadle and Tatum created several environments, each lacking only one substance. Now, instead of looking at eye color or the shapes of wings, they watched patterns of life and death.

One of the substances *Neurospora* needed to survive was vitamin B, which it had to build from raw materials in its environment. To do this, it required a particular enzyme that was missing in one of the mutant strains of mold. Did this mean that a single gene was missing, or had several things gone wrong? Patterns of heredity could tell. Beadle and Tatum mated the mutant form with normal strains of mold, and the mutation was passed along in the 1:3 proportion predicted by Mendel's laws. The mutation clearly involved one recessive gene.

They also knew that the mold lacked one specific enzyme. Beadle and Tatum had proved the principle that one gene makes one enzyme. For their accomplishments, they shared the 1958 Nobel Prize for physiology or medicine with another geneticist, Joshua Lederberg (1925–2008). The discovery was crucial because it paved the way for a new type of genetics, focused on what genes were made of and how that determined their functions.

3

Molecular Genetics: What Genes Are and How They Work (1950–1970)

As the theory of evolution triggered a great scientific and public debate in the late 19th century, chemists were engaged in a heated debate of their own concerning the difference between living and nonliving things. Were cells simply very sophisticated mixtures of substances that obeyed the laws of physics and chemistry—like minerals, seawater, or gases? Or was an extra force needed to bring nonliving matter to life—like the energy that Victor Frankenstein used animate his monster? Until 1828, most scientists believed that living substances could only arise with the addition of some sort of mystical life force, but in that year the young German chemist Friederich Wöhler (1800–82) created the first organic substance from inorganic material in the laboratory. The experiment used potassium cyanate and ammonium sulfate to produce urea. It demonstrated that some substances in human bodies could be made without a special force; perhaps all of them could be.

Today's biology takes a materialistic approach—it tries to explain living processes on the basis of chemistry and physics, without referring to mystical forces. Until the 1950s, however, the question was still very much open. The discovery of DNA's role in heredity and evolution was crucial in convincing scientists that it might be possible to explain life in materialistic terms.

PHYSICS STIMULATES NEW WAYS OF THINKING ABOUT GENES

As biologists attacked the problem of the gene, the field of physics was undergoing an amazing revolution. Scientists such as Albert Einstein, Werner Heisenberg, Max Planck, Erwin Schrödinger, and many others had invented an entirely new way of looking at matter called quantum mechanics. Previously, physicists had been thinking of electrons and other subatomic particles as tiny, solid objects orbiting an atom's nucleus the way planets move around the Sun. The new science completely changed how people thought about the relationship between energy and matter. Physicists became intrigued by what biologists were saying about the behavior of genes. Whatever these units were, they managed to organize trillions of molecules within each cell and the more than 50 trillion cells that made up a human body; perhaps they were even governed by special forces. If so, physicists wanted to learn about them.

Erwin Schrödinger put some of these questions together in a series of lectures and a popular book called *What is Life?*, published in 1944, which quickly became the inspiration for a new generation of young physicists and biologists, including a young American named James Watson. Schrödinger phrased the theme of his book this way: "How can the events in space and time which take place within the spatial boundary of a living organism be accounted for by physics and chemistry?" It was a rephrasing of the old controversy between vitalism and materialism, and Schrödinger challenged scientists to resolve the question through an analysis of life's molecules.

The book proposed that genes had to be made of a very stable substance in order to create order in the cell and survive many generations of reproduction. Whatever that substance was, it guided the construction of larger and larger structures. Schrödinger proposed two ways in which this might happen.

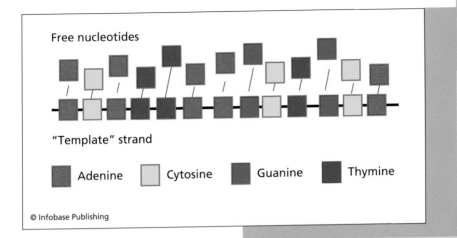

Free nucleotides

"Template" strand

■ Adenine   □ Cytosine   ■ Guanine   ■ Thymine

© Infobase Publishing

Genes either had to add on additional units that repeated over and over again, the way that a crystal grows by stacking new atoms into ever-repeating rows and columns, or they had to be chainlike molecules that grew by adding on new links. Schrödinger pointed out that even though the code had to contain all the information required to build an organism, its form might be very simple. Using the simplest possible alphabet—Morse code, which had just two "letters"—people could transmit huge amounts of information. A four-letter alphabet such as DNA might do so as well.

Hermann Muller's theoretical model of how a DNA molecule might copy itself. He reasoned that DNA might be made of two strands, with one strand holding enough information to make the second. Each nucleotide base in the strand might attract another copy of itself, which would line up free nucleotides in the right order to make a second strand. The basic principle was correct—except that bases did not attract copies of themselves. Watson and Crick discovered that each base attracted one complementary base: A binds to T, and G binds to C.

Working in Texas, Hermann Muller was concentrating on the fact that genes had to be able to create copies of themselves out of raw materials. He sketched a gene as a row of building blocks, strung together on a string. Suppose that each of the blocks was able to attract another unit of the same type. This would create a second, parallel line that contained all the elements of a new gene—the only thing missing to form a perfect copy would be to glue the new elements together into the second strand. Muller made a logical mistake when he assumed that each of the units would attract an identical unit, but, in principle, his hypothesis about the way genetic information could be copied was correct.

GENES ARE MADE OF DNA

The young German geneticist Max Delbrück (1906–81) arrived in California along with a wave of scientists escaping Germany on the eve of World War II. Just before leaving Germany, he had published an important paper called "On the Nature of Gene Mutations and Gene Structure" with a group of Berlin scientists. Known as the "Green Pamphlet" because it was bound in a green cover, it became an important inspiration for Schrödinger's book *What is Life?* After a short stay in Thomas Hunt Morgan's lab at Caltech, Delbrück moved on to Vanderbilt University in Tennessee, where he began studying *bacteriophage,* a type of virus that attacks bacteria. Many scientists had become interested in bacteriophage because some bacteria could survive infections and pass their resistance along to their offspring. Because bacteria were simpler than plant or animal cells, researchers hoped it would be easier to identify the hereditary substance that had changed to help them escape.

But very little was known about heredity in these cells. The fact that they had no visible chromosomes led some scientists to think that they reproduced using a completely different type of chemistry than plants and animals. They might not even follow the basic laws of evolution. That bacteria could adapt to the virus so fast was odd, almost as if they "learned" to survive the attacks and passed the ability on to their offspring. If so, it would be a Lamarckian type of evolution in which changes experienced by organisms during their lifetimes could enter the hereditary material.

At a physics conference in 1940, Delbrück met another immigrant, an Italian doctor-turned-bacteria-researcher named Salvador Luria (1912–91), and they spent hours talking about bacteriophage. They met at Cold Spring Harbor Laboratory in New York the following summer to carry out experiments, hoping to discover the mechanism behind bacterial heredity. Working together over the course of the next several years, they showed that bacteria were not carrying out some sort of mysterious genetic learning. Instead, they were witnessing a perfect example of natural selection in the test tube. Random mutations led to a few variants of the bacteria with immunity, and the virus wiped out cells that did not have it. The survivors passed along their mutant version of genes when they divided to make new bacteria.

This meant that bacteria evolved according to the same rules as other organisms, and their genes might be made of the same substance. The project attracted the attention of others working on the problem, including Alfred Hershey (1908–97), a bacteriologist from Michigan who paid a visit to Delbrück's lab in Tennessee. Because of the importance of their findings, Luria, Delbrück, and Hershey were awarded the Nobel Prize in physiology or medicine in 1969.

They still had not solved the riddle; instead, they had raised a troubling question. If bacteria had no chromosomes, were researchers wrong about heredity in plants and animals? Or did bacteria use some completely different type of chemistry, yet one that still obeyed Mendelian principles? The first part of the answer came from Frederick Griffith (1879–1941), a medical officer at the Ministry of Health in London. He had been studying two strains of bacteria that were very similar, trying to figure out why one of them caused severe pneumonia infections in humans and the other did not. There was only one obvious difference: The infectious, "smooth" (S) form of the bacterium built a capsule around itself, while the "rough" form (R) did not. Griffith inoculated mice with a mixture of dead S-type and live R-type bacteria. He expected that the mice would stay healthy and the bacteria would die, because he had not injected the animals with any live infectious cells. But when he drew blood, he found S-type bacteria that were alive.

Either the S type had somehow been brought back to life, or something had changed the R bacteria into the S type. If the latter were the case, it meant that R bacteria were acquiring new hereditary information. Griffith began a new round of experiments to try to find out what this transforming substance was made of. It could be that fragments of proteins from the S bacteria were somehow being absorbed into R bacteria, which then used them to build their own capsules, but Griffith had a better idea. Rather than receiving building materials to make new capsules, the bacteria might be receiving the instructions for making them. In other words, R bacteria had developed the capacity to make a new protein.

Griffith's investigations ended with his death in the Nazi bombings of London in 1941. But his work had attracted the interest of another scientist. Oswald Avery (1877–1955), a physician and researcher at the Rockefeller Institute in New York, was trying to develop a vaccine for pneumonia. That work became unnecessary because of the discovery of antibiotics, which very effectively killed pneumonia bacteria. (Today there are renewed attempts to make vaccines to kill bacteria, because so many strains have developed resistance to antibiotics.) But Avery realized that Griffith's experiments looked like the most promising way to find bacteria's hereditary material. Members of his lab purified DNA from the S type and showed that this molecule alone was able to transform the R type into infectious pneumonia bacteria. Avery cautiously proposed that in bacteria, DNA was the hereditary material and that perhaps this was true of other forms of life as well. Yet other researchers remained skeptical.

One person who believed him was Erwin Chargaff (1905–2002), an Austrian working nearby at Columbia University. In a 1971 article called "Preface to a Grammar of Biology," published in the journal *Science,* Chargaff wrote, "Avery gave us the first text of a new language, or rather he showed us where to look for it. I resolved to search for this text." If DNA was truly the language of heredity, it could not be the same in every species, so Chargaff began trying to find differences.

He knew that DNA was made of a very simple language, a four-letter alphabet of *nucleotide* bases: guanine, cytosine, adenine, and thymine (abbreviated G, C, A, and T). He started out by simply comparing how much of each base could be found in yeast cells and the tuberculosis bacterium. By chance, he had chosen two organisms with major differences in composition of their DNA. Yeast had high amounts of A and T but much lower amounts of G and C, exactly the opposite of the bacterium. Chargaff tried the same thing with other organisms and found that each had its own particular "recipe" of DNA. In humans, for example, 30.5 percent of the DNA was A, 31.8 percent was T, 17.2 percent was C, and 18.4 percent was G. The tuberculosis bacterium gave a much different picture: 15 percent A, 13.6 percent T, 34 percent C, and 37.4 percent G.

This forced many scientists to change what they had believed about the way the bases were assembled. One proposal came in 1910 from a Russian biochemist named Phoebus Levene (1869–1940), who had immigrated to the United States and was working at the Rockefeller Institute in New York. Levene proposed a hypothesis that DNA was composed of small identical units, each containing one copy of each of the four bases, which repeated over and over again. But that would mean the four bases should be present in the same amounts in every organism. Chargaff's numbers proved that this could not be the case.

If each organism had its own recipe of bases, then Avery might be right, and E. B. Wilson might have been right decades earlier: DNA might be the molecule of heredity. Chargaff noticed an extremely interesting fact: In any given organism, A and T were found in almost identical amounts; the same was true of G and C. Although he did not realize it, these numbers provided one of the most important clues as to how the DNA molecule was put together. It would not be clarified until James Watson and Francis Crick explained DNA's structure.

THE DOUBLE HELIX

By the late 1930s, biology had changed so much that it needed a new name. It was christened in a speech given in 1938 by Warren Weaver (1894–1978), director of natural sciences at the Rockefeller Foundation in New York. Seeing that the focus of the life sciences was moving toward the fundamental chemical units within cells, he said the field should be called molecular biology.

Classical genetics was still moving along as the students of Thomas Morgan established new laboratories in universities throughout the country, where they continued to work with animals and discover new genes. But many scientists were shifting their focus to how the fundamental units of cells, proteins and other molecules, carried out chemical reactions. Within just a few years of Weaver's statement, after dramatic discoveries about the nature of genes, nearly everyone in the field would consider themselves molecular biologists.

By 1950, Avery and Chargaff's work had convinced many scientists that genes were made of DNA, and laboratories across the world raced to prove it. Understanding how this molecule was put together might solve some of the questions that scientists had about genes. Chemists knew that the molecule formed a long string. They also knew that it consisted of a sugar called deoxyribose, plenty of phosphate atoms, and the four nucleotide bases. Each of these chemical building blocks had a particular shape, like a puzzle piece with sticky edges, but the chemistry was so complicated that there were too many possible answers about how they might fit together.

The details of DNA and other molecules such as proteins were too small to be seen through even the most powerful electron microscopes, so chemists were trying to understand DNA's structure by watching how other molecules changed it—a bit like ramming cars into each other to study their engines. Crystallography took another approach, turning molecules into crystals and exposing them to X-rays. This had provided some important information about the shapes of proteins; perhaps the same thing would work with DNA.

When an X-ray beam passes through an object, some of the waves collide with atoms' electrons and are scattered, or diffracted (deflected off in a new direction). William Astbury, (1898–1961), a British physicist and biologist, shone X-rays through molecules and captured the scattering patterns on photographic plates. Usually the resulting image was an unreadable smear. But if a molecule's atoms were arranged in precise, repeated patterns, waves were scattered in the same directions over and over again, creating a symmetrical pattern that hinted at the shapes of molecules. Astbury had been trying this with proteins that had formed crystals, which were ideal for X-ray studies. In some crystals, molecules are arranged in precise lattices that repeat over and over again, billions or trillions of times.

Chemists had shown that very pure DNA could be made into crystals, or pulled into fibers that also provided regular diffraction patterns if an X-ray beam hit them at the right angle. When Astbury examined DNA with X-rays, he obtained some basic information about the size and architecture of the molecule. His interpretation was that the bases fit together into flat disks, squeezed very tightly together like dinner plates stacked in a column. He could measure the diameter of the disks and the height of each plate. However, many of the details were blurred, because without knowing it, he was working with two different forms of DNA. In his images, they were superimposed.

The problem interested the great American chemist Linus Pauling (1901–94) and his laboratory at Caltech. He carried out similar experiments and proposed a structure for DNA showing the molecule as a braid of three strands organized in a helix, like a spiral staircase with three handrails. It was one of the few times Pauling was wrong. Considered to be one of the greatest chemists of the 20th century, he had made great advances in the use of crystallography to understand the building plans

The American chemist Linus Pauling, who had done brilliant work with protein structures, was one of many scientists to propose a flawed model of the structure of DNA. *(Ava Helen and Linus Pauling Papers, Oregon State University Library)*

of proteins; this work earned him a Nobel Prize in chemistry in 1954. Eight years later, he became only the second person in history to win a second prize in a different category (the other was Marie Curie). This time it was the 1962 Nobel Peace Prize for his efforts to stop the testing of nuclear weapons aboveground.

Pauling had become interested in political issues thanks to the efforts of his wife, Ava, a peace activist and human rights advocate. There was a cost; in 1952, he had been denounced as a communist before Senator Joseph McCarthy's committee in the U.S. Congress. This had an impact on his scientific career because that same year, the government refused to grant Pauling a visa to attend a scientific meeting of the Royal Society in London. One of his colleagues, Robert Corey, went instead. During the trip, Corey met with a young researcher named Rosalind Franklin (1920–58), who was using crystallography to investigate DNA. It is hard to tell what might have happened had Pauling attended the meeting, but he might have obtained data that would have helped him create an accurate model of DNA. Franklin's work was about to play a crucial role in figuring out the molecule's structure.

Another incorrect model had just been proposed by the British scientist Francis Crick (1916–2004) and his young American partner James Watson (1928–), working in Cambridge, England. Watson had obtained his Ph.D. at the age of 22, studying on bacteriophage at the University of Indiana, and had come to Cambridge determined to solve the riddle of DNA's structure. He was now 23, and Crick was 35, but the two men quickly recognized each other as two of the brightest people on campus and hit it off. They had a lot of catching up to do when it came to DNA; neither was an expert in chemistry. Their first diagram of the molecule was so wrong that it embarrassed their boss, Lawrence Bragg, and he ordered them to stop working on it.

Meanwhile, Rosalind Franklin, an hour away by train in the laboratory of Maurice Wilkins in London, had solved a major problem regarding the X-ray images of DNA. She had figured out that DNA came in two forms—A and B—a "dry" and a "wet" form. Under humid conditions, more hydrogen atoms were packed into the molecule, and that changed its shape. She realized that in the images taken by Astbury and Pauling, the forms were superimposed, and the image was blurred. Using only the B form, Franklin obtained the sharpest-ever images of DNA. She began trying to interpret what this meant about its structure but interrupted the work to go on vacation. While she was gone, Wilkins showed some of her X-ray images to Watson. One look at a photograph of the B form was enough to convince them that DNA formed a double helix.

The problem that now faced Watson and Crick was like trying to solve one of those wooden puzzles in which oddly shaped pieces have to be fit together to form a geometric shape. In this case, the shape that had to be built was a helix, and the pieces were sugars, phosphates, and bases. Watson made cardboard cutouts in the shape of the four bases and began working on the puzzle. No matter how he attached them to each other, something always bulged outside the helix. He was stuck until his office mate—ironically a former student of their competitor, Linus Pauling—told him that bases existed in two different chemical forms, with slightly different shapes. Following the chemistry textbooks, Watson had been using a form with an extra oxygen atom. His office mate told him that the textbooks were wrong, so Watson remade the shapes based on the type without the oxygen. He was idly fitting them together when he had a sudden revelation: When A snapped onto T, it had almost exactly the same size as G fit to C. Fit together, their size matched the dimensions of the helix in Rosalind Franklin's X-ray photographs.

Crick came in, and Watson showed him what he had discovered. They immediately realized what it meant: The "steps" of the DNA spiral staircase were the bases, rather than the sugars. Each step had either an A combined with T, or a G with a C. The steps were connected by winding "rails" of deoxyribose sugars (the backbone). Between each of the steps, there was a slight twist, making the whole structure into a helix rather than a straight column. They quickly wrote a paper called "A Structure for Deoxyribose Nucleic Acid" and submitted it to the journal *Nature*. It was published three weeks later—an amazingly short time, given the fact that it first had to be read and commented on by experts.

Francis Crick and James Watson in front of their DNA model *(A. Barrington Brown/Science Photo Library)*

This brief article would revolutionize biology because the molecule's building plan provided immediate insights into its behavior. The new model explained Chargaff's discovery that A and T occur in identical amounts in an organism, as do G and C. And the way the bases came in pairs revealed how DNA might copy itself. If the two strands of DNA were split apart, each base would attract and link up to just the right partner nucleotide, creating a second strand. The article even suggested a way that mutations could occur, in spite of the fact that bases formed regular pairs. In rare cases, hydrogen atoms might bind differently to a base, slightly changing its shape. As one strand was copied, it might then attach to the wrong base.

Another important point was that any *sequence*—any possible "spelling" of the four bases—formed the same shape. A long string made up only of As joined to Ts would create the same double helix as a sequence consisting only of G-C pairs. Each organism could have its own DNA sequence; a language with four letters was rich enough to create all the diversity of life on Earth. Evolution was built on a single scaffold.

With this single, powerful image, some of the most important questions about genes, cell replication, and evolution were resolved, all at once. The double helix was the last blow to the idea that a special force was needed to explain life—from that point on, the goal was to explain what happened in cells and organisms in terms of materialist physics and chemistry.

An article containing the X-ray data from Franklin and Wilkins appeared in the same issue of *Nature* as the Watson-Crick article. Nine years later, Watson, Crick, and Wilkins were awarded the Nobel Prize in medicine or physiology for their discoveries.

There has been a debate about the fairness of the award ever since, particularly because many historians believe that Franklin's data was crucial to solving the puzzle. Yet the prize can only be given to three people in one category in a single year, and only to living people. Franklin had died of cancer in 1958, possibly as a result of her long work with X-rays. The idea that she was treated unfairly is based on claims that there were political and personal reasons behind the decision. Her relationship with Wilkins had always been strained. And the degree to which Watson and Crick had been inspired by her data only became clear later, when Watson told his version of the discovery in a book called *The Double Helix*.

RNA IS THE MESSENGER

Within a short time, so much evidence accumulated for Watson and Crick's model—collected by Franklin, Wilkins, and others—that even the sharpest critics had to admit that the question of the chemical basis of heredity had been solved. Yet this was clearly only a beginning. Whereas Beadle and Tatum had proven that genes were responsible for making proteins, no one understood how the information in DNA could become transformed into a completely different type of molecule. This question would preoccupy scientists for the next 15 years.

Biochemists already knew how the cell made most of its components. They had worked out the steps involved in making sugars, fats, the single DNA bases, vitamins, and the single building blocks of proteins, called *amino acids.* But how could cells assemble hundreds or thousands of amino acids into the same type of protein, over and over again, always the same way?

As early as 1942, a Belgian researcher named Jean Brachet (1909–98) claimed that molecules called ribonucleic acids (RNAs) played a role in protein building. RNAs were made of nucleic acids, like DNA, but they contained a base called uracil (U) instead of thymine (T). And instead of being connected by deoxyribose sugar handrails, like DNA, the nucleotides in RNA were linked by another type of sugar, called ribose. Another difference was that RNA did not usually form double strands, the way DNA does, because the cell usually does not make RNAs with complementary sequences.

Brachet thought that RNA helped transmit information from genes to proteins because of its location in the cell. DNA was stored in the nucleus; most proteins were found in the surrounding regions of the cell (the cytoplasm). Most RNAs could be found there, too, but some could be seen in a pocketlike region of the nucleus called the nucleolus. Unlike DNA, RNA seemed to be able to travel between the two major compartments of the cell.

Cyril Hinshelwood (1897–1967), a chemist at Oxford University, reasoned that the cell might make an RNA pattern molecule for each protein, like an instruction manual to assemble amino acids in the proper order. If that were the case, the cell had a method of translating a four-letter language (the four nucleotides) into a 20-letter alphabet (the 20 amino acids that make up proteins). Essentially, he said, it was a code-breaking problem. The British had honed their code-breaking skills during World War II, only a decade in the past, with their efforts to break the incredibly complex Enigma codes used by the German military. (Mathematicians, physicists, and a few biologists had been involved in the code-breaking activity. Others were drawn into the war effort in some strange ways. Francis Crick had built mine detectors until a bomb blew up one of his devices. Max Perutz, one of Crick's colleagues and an expert in proteins, had been drafted for a crazy scheme to manufacture artificial icebergs to use as floating runways for airplanes.)

Even before DNA's structure had been solved, an American chemist named Alexander Dounce (1910–97) of the University of Rochester in New York had proposed a theory regarding the code problem. He did not know whether DNA or RNA contained the genetic code, but it made no difference mathematically, because each type of molecule was written in a four-letter language. Translating a four-letter system into 20 amino acids would require at least three bases to spell one amino acid. Two letters were not enough, because there were only 16 possible ways that two bases could be arranged in two-letter words (4^2): AA, AG, AC, AT, GG, GA, GC, etc. But three-letter words (4^3) would permit 64 possible spellings. It did not matter that more words were possible than cells actually used; several different spellings might be used to create the same amino acid.

Hinshelwood and other scientists had been focusing on RNA as the patterning molecule, but the breakthrough in DNA made some scientists think that genes might directly create proteins themselves, without the need for a middleman. One proponent of the idea was George Gamow (1904–68), a Russian physicist famous for proposing the "big bang" theory of the origin of the universe, who had defected to the United States. In 1953, he wrote a letter to Watson suggesting that amino acids might attach themselves directly to the surface of the double helix, where they could be strung together. Gamow's hypothesis turned out to be wrong, but it aroused Francis Crick's interest. In 1958, Crick gave a series of talks and papers mapping out a strategy to find how the information in genes was transformed into proteins, which would keep scientists busy for the next 15 years.

The plan revolved around an idea that Crick had developed with Watson. He called it the central dogma of molecular biology: "DNA makes RNA makes proteins." This concept had several implications, Crick said. First, it confirmed the one gene-one enzyme principle of Beadle and Tatum. As genes carried hereditary information, they encoded the recipes for protein molecules that carried out most cell functions. A second point was that genes were strings that had a beginning and ran to the end in a particular order, like words in a sentence. Since proteins were also strings, there had to be a key in order to translate one code into the other. The main goal for molecular biologists should be to discover the key and to find the molecules responsible for the translation.

First, the sequence of a gene was transcribed into an RNA molecule, which had a similar chemistry. Then the RNA was transported out of the cell nucleus to the cytoplasm, where it was translated into protein. The RNA was a long string of letters—how did the cell know where one word ended and the next began? AGGAGGAGGAGG could be read as AGG AGG AGG AGG, or GGA GGA GGA GGA, or GAG GAG GAG GAG, depending on where the borders between the letters fell. This would cause problems, because AGG might spell a different amino acid than GGA. If one were to move all the spaces in a normal sentence ("The man ate his old red car") and put them in at the wrong places ("Th ema nat ehi sol dre dca r"), most of the words would be nonsense.

The cell had to have some sort of system to prevent this from happening. Perhaps there was a marking system that placed something like commas between the words. Or there might be a signal telling the cell where to start reading, after which it would simply read the first three letters, then the next three, and on to the end.

In either case, Crick suggested, the cell might hold something like a dictionary that helped it distinguish between real words and nonsense. There had to be something like chemical adapters, able to fit a three-pronged plug into an outlet with only one hole. Crick proposed that to do this, the cell needed 20 different adapter molecules, along with 20 enzymes able to actually do the plugging. Chemists began looking for such molecules and quickly found them. Robert Holley (1922–1993), a biochemist at Cornell University, found the first adapter, an RNA molecule he called transfer RNA (tRNA).

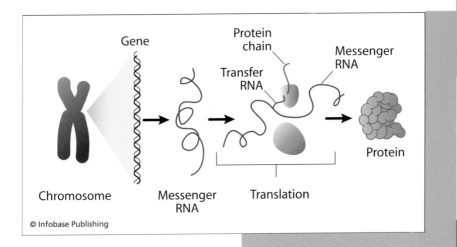

Gene

Protein chain

Transfer RNA

Messenger RNA

Protein

Chromosome

Messenger RNA

Translation

© Infobase Publishing

Meanwhile, scientists were beginning to figure out the RNA spellings of particular amino acids. Marshall Nirenberg (1927–), of the National Institutes of Health, found the first codon while working with an artificial RNA that consisted entirely of the nucleotide uracil (U). When he put this into a test tube with protein-building molecules, they read it and built a protein that consisted of only one type of amino acid: phenylalanine. This meant that the codon for phenylalanine had to be spelled "UUU." When Nirenberg made his techniques public, other scientists followed his lead. They hoped to create or find similar molecules with repeated, identical codons, but chemistry was not far enough along to build made-to-order RNAs.

Information in genes is transcribed into a chemically similar molecule called messenger RNA, which is then translated into a protein by the ribosome. Other molecules regulate each step of the process, allowing the cell to control when and where particular proteins are produced.

Then Marianne Grunberg-Manago (1921–), working at the New York University School of Medicine with a Spanish-American chemist named Severo Ochoa (1905–93), discovered an enzyme in bacteria that joined random nucleotides into RNA-like strings. Putting this molecule into Nirenberg's test-tube system produced a wide variety of proteins. There was no way to directly see what types of RNAs the enzyme made, but the scientists came up with a clever way to make good guesses. By adding different proportions of nucleotides, they could estimate how often certain combinations should appear. This was a bit like putting Scrabble letters into a bag and drawing out three at a time, without looking. Putting in specific quantities of certain letters—10 times A, 15 times B, four times C, etc.—would make it possible to guess how often certain combinations should be drawn. Counting the most frequent amino acids that were made allowed the scientists to guess more of the spellings. Ochoa was awarded the 1959 Nobel Prize in physiology or medicine for the work.

The method was still limited because there was no way to influence the random spellings of the RNAs. In the early 1960s, Har Khorana (1922–), a native of India working at the University of Wisconsin, developed new techniques that allowed him to create small, artificial RNAs with the spellings he desired. Within a few years, this strategy led to the working out of the entire code. The result was another Nobel Prize in physiology or medicine, shared by Holley, Khorana, and Nirenberg in 1968.

Crick's hypothesis that each amino acid matches a particular three-letter codon was confirmed. Four nucleotides could be combined to make 64 possible

In a lecture series in 1958, Francis Crick announced the "central dogma" of molecular biology: "DNA makes RNA makes protein," laying out a road map for the next 15 years of research in molecular biology. *(Wellcome Library for the History and Understanding of Medicine, Francis Harry Compton Crick Papers)*

three-letter codons. This is more than an organism needs, but some amino acids have two or more different spellings. Leucine has the most; it can be spelled from six different codons. Three combinations are nonsense that do not spell amino acids at all. They are called *stop codons* because when the cellular machinery sees one, it breaks off the synthesis of a protein.

THE ARCHITECTURE OF GENES

The building plan of DNA showed how it might be copied, but almost nothing was known about the behavior of genes. Many of the details were worked out by the French researchers François Jacob (1920–), Jacques Monod (1910–76), and their colleagues at the Pasteur Institute in Paris. Jacob and Monod complemented each other perfectly: Jacob seemed to have an instinctive grasp of how the cell worked, and Monod was a genius at breaking down complex problems into steps that could be tackled in experiments. With their colleagues at Pasteur, they made discovery after discovery about genes.

Evidence suggested that in addition to passing on genes to their offspring, bacteria could sometimes pass them to their "brothers and sisters," a problem that interested Jacob and his colleague Élie Wollman (1917–). Some scientists claimed that this happened because certain types of bacteria could mate, but the idea was still very controversial. Jacob proved it with the help of an ordinary kitchen blender, purchased as a surprise for his wife during a trip to the United States. His wife was "disgusted" by it, saying that such a device had no business in a French kitchen, so he took it to the laboratory, where it soon played a role in a famous experiment.

Jacob and Wollman grew male and female bacteria in isolation from each other. To discover whether the cells could mate, they had to prove that a gene moved between them, so they used females with a defective form of a gene called lac B. The function of this molecule was to help bacteria break down milk sugar into two parts, galactose and lactose. Bacteria could survive without the gene, so it would be ideal for the test. The male version of the gene was healthy. When the scientists mixed cells of the two sexes, they discovered that the females acquired the healthy gene from the males, proving that the types were really mating.

This process, called conjugation, was very slow. It might take up to two hours, and that led to another interesting discovery. If mating was interrupted in the middle, only some genes were transferred. This was quite different than the all-or-nothing fusion of chromosomes that happens when plants or animals mate. With bacteria, however, the longer mating went on, the more genes were transferred, always in the same order. The situation was like people in a cinema leaving in the middle of a bad movie—they would see the same story, and events in the same order, but only those who stayed for the whole two hours would see the whole thing.

Jacob discussed the findings with Monod and came up with the "spaghetti hypothesis" of the transfer of bacterial genes. A bacteria's chromosome might be like a long piece of spaghetti, with genes in a particular order, which was slowly threaded from male to female. If mating was interrupted (by shaking the bacteria apart in the blender), the strand broke off, and no more genes were transferred. If the idea was correct, there had to be some type of physical transfer of the "spaghetti," and maybe it could be seen under the electron microscope. A close look soon revealed that a tiny bridge formed as the cells mated. Stringlike DNA was being transferred across the bridge.

Because the process was so slow, Jacob and Wollman could interrupt mating at precise times. This broke the chromosome at specific places, and it allowed them to make an exact map of the positions of genes on the chromosome. This revealed another strange fact: The genes always remained in the same order, but sometimes the strand of spaghetti began in a different place. If there were seven genes, for example, they were sometimes found in the order ABCDEFG and sometimes CDEFGAB. Jacob had a brilliant insight: Bacteria had chromosomes shaped like circles rather than straight rods. To prove it, once again the researchers turned to the microscope. They discovered ringlike structures to which they gave the name *plasmids.*

During mating, the damaged female lac B gene was somehow replaced with the working male version—how did this happen? Over the next few years, researchers learned that the replacement had to do with the way bacteria repaired damage to their DNA. Molecules in the female were continually on the lookout for loose genes that might have been lost. Upon finding one, they plugged it back into the chromosome and kicked out anything that might be occupying its place. This process of repair, called transformation, was a key to the way bacteria developed resistance to antibiotics. Two decades later, researchers would learn to adapt the process to plants and animals, launching the age of genetic engineering.

ON-OFF SWITCHES FOR GENES

Cells do not constantly make proteins from all of their genes; only about 20 percent of human genes, for example, are active in a particular cell at any one time. This implied that genes had to have some sort of on-off switch. Monod was sure that other molecules, such as proteins, were doing this job, but how did they work? He reasoned that the problem ought to be easy to study in bacteria. To adjust to changes in the food supply or other changes in the environment, bacteria required different molecules. That meant activating different sets of genes, thereby producing new RNAs and proteins.

Monod originally supposed that the normal state of a gene was off until a protein threw its switch. Soon experiments convinced him that he had it backward. Left to their own devices, genes would be stuck in the on mode, producing proteins all the time. Silent genes were being held in check by brakes—control molecules that Monod and Jacob called repressors. To create a protein, it was necessary to release the brake.

This work also shed light on RNAs. Scientists knew that the creation of proteins required huge clusters of molecules called ribosomes. One theory at the time was that the cell built a special ribosome for each protein that had to be made. Monod and Jacob were skeptical because this would obviously be an enormous job for the cell, and their experiments showed that genes could start turning out proteins quickly. The Paris group presented their ideas to Francis Crick and Sydney Brenner (1927–), a South African geneticist working in Cambridge, England. They developed the correct interpretation of ribosomes as huge, generic "reading" machines, able to read any RNA and to produce thousands of different types of proteins.

Each gene likely had a unique repressor that had to be able to find the right target on DNA, where it could exert its influence on a nearby gene. Jacob and Monod named these targets operators and thought of a way to find them. If an operator experienced a mutation, a repressor might no longer be able to find it or bind to it. Operators were not genes themselves, because they did not hold the information necessary to create proteins; instead, their function was to receive a signal—a molecule that acted like a radio transmission. If the receiver was broken, no signal would arrive, and a gene under its control would not behave properly. Experiments with bacteria showed that this was exactly what happened when a mutation changed an operator. In 1960, the scientists drew their ideas together into a new concept called the operon: a structure in chromosomes that contained both genes and their controlling regions.

Soon, they saw that operons had even more parts. At the beginning came a *promoter* region whose function was to attract an RNA *polymerase,* the molecule that actually built the RNA. (A polymerase is a general name for enzymes that build other molecules by gluing together smaller units.) Promoters for different genes had particular features, and their chemistry might change depending on what was happening in the cell. When a lot of a particular protein was needed, the promoter for its gene became chemically very attractive to RNA polymerases. Genes needed in small quantities had less attractive promoters.

The next segment of the operon contained the operator. If a repressor sat here, it would act as an obstacle as the RNA polymerase tried to slide down the gene. The polymerase would be thrown off track, and no RNA or protein would be produced from the gene. Monod only had to make one more hypothesis to explain how a protein could activate a gene. It might latch onto a repressor molecule, which then let go of the DNA, giving polymerases free access to the gene.

THE FLOW OF INFORMATION FROM GENE TO PROTEIN IN COMPLEX ORGANISMS

"DNA makes RNA makes proteins," the road map announced by Watson and Crick, had now been proven; the new goal was to work out the details of each step. Scientists discovered how the DNA double helix was teased apart, creating separate strands that could be copied by other molecules. RNA polymerases read genes and transcribed them into RNA molecules. Ribosomes then attached themselves to RNAs and translated them into proteins with the help of transfer RNAs and adapter molecules.

These basic elements in the transformation of genetic information into proteins evolved in a cell that lived at least two billion years ago; it has been inherited by all organisms alive on Earth today. At least 1.5 billion years ago, cells took on different evolutionary routes as fundamental biological processes were added and developed in different ways. Eukaryotes, the branch of life containing yeast, other molds, plants, and animals, developed an additional internal compartment called the nucleus. This created a barrier that could be used to prevent molecules from reaching DNA; it also had to be crossed when RNAs left the nucleus. The nucleus permitted the evolution of many new steps in the pathway of gene to protein. Uncovering and investigating these steps has been a major focus of biology since the 1970s.

One discovery was that a single gene could be used to create several different proteins. Thus, while scientists currently estimate that the human genome holds somewhere from 23,000 to 26,000 genes, it may produce dozens or hundreds of times as many proteins. This comes from the boxcarlike nature of genes in plants and animals. Most are interrupted by many extra regions called *introns* that do not hold protein-encoding information. They have to be removed before an RNA is allowed to leave the nucleus. Introns are first transcribed into RNA, like the rest of the gene, but are then removed in a cut-and-paste operation called *splicing.*

One way to imagine this is to think of RNA as a train with many boxcars. Introns would be like empty cars, scattered between loaded ones, which have to be unlinked and removed before the train pulls out of the station. Cars containing protein-encoding information, called *exons,* are then linked back together and sent along to their destination. Sometimes trains are assembled for special customers, leaving out exons as well, in a process called *alternative splicing.* Thus, very complex genes (with a lot of exons) can be used to produce hundreds or even thousands of particular types of protein molecules by mixing and matching different sets of cars. When splicing was discovered, this was thought to be an exotic, rare process; now it is known that on the average, human genes hold 8.4 introns per gene—all of which have to be removed through splicing. The record holder for splicing seems to be a gene called Dscam, found in the fruit fly, which can potentially be used as a template for 38,000 differently constructed proteins.

The evolution of alternative splicing created new diversity in organisms, because it enlarged the number of proteins that cells could produce without the evolution of new genes. Some curious biological processes depend on it. For example, alternative splicing of a complex gene helps some species of birds distinguish between the different tones of songs.

Alternative splicing also gave cells new ways to fine-tune the use of their genes. In the early days of molecular biology, most scientists thought that decisions about producing proteins were made at the first step by keeping a gene switched off with a repressor, or not attracting RNA polymerases to its promoter. Today, several other modes of control are known, often involving RNAs. One system notices some RNAs that have not been properly spliced and traps them in the nucleus. This means they cannot reach the cytoplasm to be translated into proteins. Another system called *nonsense-mediated mRNA decay* (NMD) catches RNAs in which mutations have turned information to nonsense. These molecules are usually recognized and broken

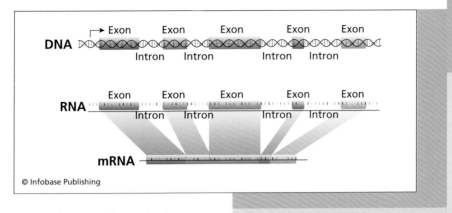

Exon Exon Exon Exon Exon

DNA

Intron Intron Intron Intron

Exon Exon Exon Exon Exon

RNA

Intron Intron Intron Intron

mRNA

© Infobase Publishing

down before they do harm. Control can also happen later; RNAs may be put on hold until they are needed or until they reach particular locations in the cell. This is often accomplished by tagging an RNA with proteins that block ribosomes from reading the molecule.

Within a period of 20 years, the revolution that had begun by linking genes to DNA had solved major questions about how genetic information guided the life of cells and organisms. The discovery that DNA had a regular chemical structure and that it obeyed the normal rules of chemistry meant that no special forces were needed to explain life. Most scientists believed that cells could be regarded as machines, despite their incredible complexity. By explaining how DNA could be copied, how mistakes could arise, and how mutations could affect the cell (by altering the makeup and behavior of proteins), the discoveries brought genetics and evolution back together after decades of debate about whether the two systems were compatible with each other.

Most genes consist of a mixture of protein-encoding regions called exons and long stretches of noncoding material called introns that have to be removed. This happens in a series of steps: First the entire sequence is transcribed into an RNA molecule. Then the introns are eliminated in a process called splicing, leaving a finished messenger RNA. In some cases, particular exons are also removed, which means that the same gene can produce several mRNAs and thus different forms of a protein.

4

The Rise of Genetic Engineering (1970–1990)

By the end of the 20th century, scientists were routinely transplanting genes between organisms, shutting them down to study their functions, and manipulating them in many other ways. These technologies were being applied to the production of new foods and medicines, and many other uses could be foreseen. Bacteria might be specially engineered to clean up oil spills or other types of pollution, artificial viruses might deliver healthy DNA to people suffering from genetic illnesses, and the body's immune system could be reprogrammed to fight off cancer and other diseases. At the same time, genetic engineering is constantly presented as a scary topic in the headlines. A few examples from 2007 were "Franken-Broccoli? The GM Seed Giants Lumber into the Veggie Patch," from the December 19 issue of *Grist* magazine; "Attack of the Mutant Biotech Rice," from the July 9 issue of *Fortune*; and "Genetically Engineered Organisms Invade Our Planet," printed in the *Epoch Times* of March 12. What headlines usually fail to capture is the fact that genetic engineering has been a crucial tool in answering fundamental questions about life.

Genetic engineering was made possible by the development of new forms of biotechnology. This chapter describes how work carried out between the 1970s and 1990s produced an extremely sophisticated toolbox that researchers now routinely use to manipulate and intervene in the hereditary material of organisms ranging from bacteria to complex animals.

RECOMBINANT DNA

In the late 1950s, a Swiss scientist named Werner Arber (1929–   ) finished a Ph.D. dissertation on the genetics of phages, the bacteria-infecting viruses that had been so helpful in understanding genes. He set off for the United States, where he spent several months in the laboratories of Joshua Lederberg at Stanford University and Salvador Luria at MIT, both of whom were doing groundbreaking work on the viruses. The fact that bacteria did not always succumb to an attack had interested scientists from Delbrück and Luria to Jacob and Monod, and when Arber returned to Switzerland, he began pursuing this area in his own laboratory. Some bacteria possessed proteins that partially protected them from the virus; Luria had called them restriction enzymes because they were able to restrict infections in cultures of bacteria. Now Arber discovered how this happened. For the virus to reproduce and survive, it had to insert its DNA into bacteria and prompt the new host to copy it. Restriction enzymes recognized that the DNA was foreign and attacked it by chopping it into small fragments.

By the 1970s, Hamilton Smith (1931–   ), at Johns Hopkins University, isolated one of these enzymes from the bacterium *E. coli* and worked out many of the details of its functions. The protein Smith found, called a type II restriction enzyme, scanned DNA until it found a particular sequence, and then it made a cut. This sequence is usually found only in viruses, not in the bacteria's own DNA, which allows the enzyme to attack foreign molecules without damaging its own cell.

Smith also discovered something interesting about the way the cut was made. It did not produce a clean cut of the two strands of DNA, like sawing off the top of a wooden ladder by making a straight cut across the two rungs. Instead, the

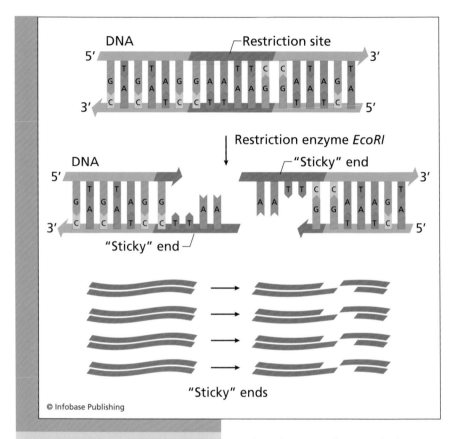

DNA   Restriction site

Restriction enzyme *EcoRI*

DNA   "Sticky" end

"Sticky" end

"Sticky" ends

© Infobase Publishing

Restriction enzymes are DNA-cutting molecules from bacteria that are used as tools in genetic engineering. The enzymes recognize particular DNA sequences and break the strand, leaving two "sticky" ends. Other molecules recognize that the broken ends match and can mend them. If researchers create an artificial molecule with broken, sticky ends that match such breaks, the repair molecules may paste it into an organism's genome.

break was diagonal, leaving one long handrail with half-rungs (the bases on one side) hanging over. This loose side is chemically sticky, and if it encounters another molecule with a similar loose end, with complementary bases, the two may join. The other side of the ladder has such a complementary sequence, which means that the broken ends can be rejoined if they are brought together. Bacteria contain another type of molecule, called a *ligase,* that can mend the cuts. Since the DNA strand is sometimes broken by mistake, ligases have an important function: They can repair damage.

Understanding how such cuts were made and how the ends were rejoined was an extremely important step on the road to genetic engineering. A restriction enzyme could be used to make a break in a chromosome. Suppose that a researcher adds loose tails with the right sequences on the ends of a gene. If this molecule is inserted into bacteria, ligases may paste it into the break. This technique, called DNA recombination, was used for the first time by two researchers at Stanford University, Janet Mertz and Ronald Davis, in 1972.

A year later, two of their colleagues, Stanley Cohen and Annie Chang, worked with Herbert Boyer of the University of California, San Francisco, to move a gene from one species to another. They combined genetic material from a virus and a bacteria and inserted it into another bacteria. The artificial gene was taken up by the cell and used to create a foreign protein. Ironically, to transplant genes across species, the scientists were making use of molecules that had almost certainly evolved to prevent foreign DNA from invading cells. These accomplishments marked the beginning of the age of genetic engineering.

MOLECULAR CLONING AND USING BACTERIA AS DRUG FACTORIES

In the popular media, *cloning* usually refers to copying entire humans or organisms. For molecular biologists, cloning usually means to copy genes or larger segments of DNA, often in bacteria. This is one of the most important tools in today's biology and medicine. Bacteria divide at a tremendous rate. With the invention of genetic engineering, it was immediately obvious that the cells might be used to turn out high quantities of useful molecules.

Molecular cloning requires inserting artificial genes into bacteria or other cells that reproduce the DNA as they divide. In the 1970s, scientists developed two main methods of insertion. One relied on ligases to put the genes into phages. The other

"Natural" Genetic Engineering

Sometimes nature manages feats of genetic "engineering" on its own—in other words, a gene is captured from one species and transplanted to another without the help of scientists. This begins when a virus or bacteria living inside a cell accidentally kidnaps genes or RNA from the host. When it infects another species, it may deposit the foreign gene there. On rare occasions, this molecule enters the hereditary material of the new host and is passed along to its offspring. This process, called horizontal gene transfer (HGT), is probably very rare, but over the course of evolution, it has happened many times. It is most common between species of bacteria, which easily absorb and integrate foreign DNA.

Viruses have also been responsible for HGTs. A virus is like a hijacker that forces the cell to give up its own activities to produce hundreds or thousands of copies of the invader. Each component of the virus has to be reproduced. This usually happens in different places in the host cell; at the end, the pieces are brought together for assembly into new infectious viruses. During the packing, genes or other molecules from the host might get mixed in. Usually this renders the virus harmless, because it is such a minimal form of life, tiny and precisely folded, with little room to spare. Inserting extra baggage usually yields an improperly built structure that cannot infect another cell. Occasionally, however, a virus may carry something extra along. In this way, viruses have shuffled DNA and RNA between species. It is impossible to know how often this has happened; some researchers believe that early in the history of the planet, it played an important role in evolution.

Many viruses bring along RNA and protein tools that prompt cells to replicate them. Retroviruses go a step further: They sneak their own genes into the genome of bacteria, plants, or animals. The foreign genes may lurk there for a long time, like a Trojan horse, waiting for the right conditions to begin making raw material to form new viruses. Thus HIV, the AIDS virus, is able to hide in human immune system cells for years after it integrates key parts of its recipe into the genetic code of cells. HIV infects blood cells, so it is not passed along to an infected person's children in the DNA of a sperm or egg. (It can be passed between mother and child through an exchange of blood during birth.) But some retroviruses have worked their way into reproductive cells. A well-known example involves a retrovirus that causes leukemia. Long ago the virus infected a cat; its DNA was integrated into that of the host and was passed down along with the cat's own genes from then on. Other retroviruses that infected ancient animals were so potent that artifacts of their DNA can be seen scattered throughout the genomes of humans, fish, and other animals.

Bacteria have also carried out HGTs. In 2004, a study by British researcher Toby Gibson at the European Molecular Biology Laboratory in Germany showed that bacteria had captured a gene called alpha-macroglobulin, probably from animals. Normally, this molecule helps hosts fight off pathogens; capturing and integrating the molecule might help the bacteria to evade cell defenses.

In an article published in the September 21, 2007, issue of *Science* magazine, Julie Hotopp and Hervé Tettelin of the J. Craig Venter Institute in Rockville, Maryland, with Michael Clark of the University of Rochester (New York), and colleagues from five other universities, discovered widespread examples of HGT in insects. The scientists focused on a bacteria called *Wolbachia* that long ago

(continues)

(continued)

infected insects and now lives as an endosymbiant—a permanent resident—in many insect species. It is estimated that the organism has infected more than 20 percent of insects and many species of parasitic worms (such as tapeworms) throughout the world. The organism has so thoroughly integrated itself into tapeworm cells that eliminating it often leads to their death or makes them sterile. In an article published in *The Lancet* in 2005, Mark Taylor of the Liverpool School of Tropical Medicine in Great Britain showed that a drug that killed *Wolbachia* bacteria was a more effective treatment for tapeworms in humans than drugs that directly attacked the worms.

Hotopp's study analyzed the genomes of worms and insects in search of evidence that the animals' cells had adopted *Wolbachia* genes into their own genomes. The scientists found that flies had adopted nearly the entire bacterial genome into their own genetic material in the nucleus. Many other species had integrated large parts of the genome. There have probably been even more HGTs than the study reveals, because when bacterial genes are found by genome projects, they are routinely ignored.

method uses plasmids, the circular chromosomes that François Jacob and Jacques Monod discovered in bacteria, described in chapter 3. These carry their own sets of tools and instructions to replicate and can copy themselves thousands of times inside a cell, independently of whether the bacteria itself is dividing. They are also replicated when the cell divides, so a colony of bacteria can contain a huge amount of a protein made from a particular gene.

Another feature of plasmids made them potentially more useful than phage DNA. Bacteria absorb phage genes from the

(Scientists often assume they are contaminations from infections rather than parts of the genome itself.)

Horizontal gene transfers have made things difficult for researchers who would like to read the history of evolution from the genetic code of today's organisms—a process that is a bit like going to the library to track the evolution of Latin into Italian, French, Spanish, and other modern languages. If there had never been any HGTs, then the only source of an organism's DNA would be its direct parents, and reconstructing a family tree from its genes would be relatively simple. HGTs make the task much more difficult, just as the import of large numbers of foreign words can confuse someone trying to study the history of a language. Building a tree of life is especially difficult when looking far back into the history of life, where HGTs may have been much more common. Some researchers even suggest that DNA became the only form of heredity on Earth because at an early stage in life, viruses commonly carried it between unicellular species that might have been using RNA to pass along hereditary information. DNA was such a stable molecule that over time, it won out over other types of heredity through natural selection.

environment, so a single cell might soak up many different genes and insert them into the genome in different places. It is hard to predict how the new genes will be used or whether they will be copied efficiently; it may also be difficult to extract the copies later. However, bacteria obtain plasmids through transfers from neighboring bacteria, and they accept only a single one. This makes it easier to ensure that each cell in a colony of bacteria has the same plasmid, and it can be extracted in very pure form.

Cloning a gene has become routine in the age of genomes; researchers now have complete maps showing where a target gene sits within an organism's DNA. But a few decades ago, this was rarely known, and the only way to find and clone a particular gene was to pull down the complete "library" of the genome. Making a single clone often required years of work without a guarantee of success. But the payoff could be huge: A single liter of bacterial cultures could produce ten times as many copies of a gene as could be found in the 100 trillion cells in a human body.

Creating a genomic library required breaking the entire human genome down into fragments using a restriction enzyme. This yielded a genomic library of fragments, one of which contained the human gene for the *insulin* hormone. The same enzyme was used to cut open plasmids, making plasmids with ends that matched the cuts in these genes. The fact that the ends were the same meant that the human gene could be glued into these breaks by a ligase to create a *bacterial artificial chromosome* (BAC).

The technique was not perfect; it produced a lot of plasmids that did not contain the gene, so another step was required to select bacteria in which the process was successful. One method involves adding even more information to the BAC: a gene that makes the bacteria resistant to antibiotics. When the bacteria are grown in cell cultures, antibiotics are applied, and they kill off any bacteria without a successful plasmid. The next step is to find colonies that contain only the gene of interest and not extra genes or other DNA; this is done by attaching a radioactive marker to the gene or looking for colonies that use it to produce proteins.

Researchers quickly discovered that bacteria did not necessarily produce useful forms of the molecules of humans or other complex organisms. Chapter 3 recounts how evolution added many steps to the relatively simple gene pathway used by bacteria and the earliest cells to synthesize proteins. For example, nearly all human genes contain large regions called introns that do not contain protein-encoding information and must be removed through splicing. But bacteria have neither introns nor the molecular machines needed to remove them. So the RNAs made from bacterial genes are direct readouts of the DNA sequence; nothing has to be removed or spliced. Simply putting copies of the complete human gene into plasmids would create RNA with extra information that would block its translation into proteins. So a version of the gene had to be made without the introns.

One method of doing so is to start with a library of complementary DNA—cDNA. Such libraries are different from the genomic libraries described above, because they are made from RNA molecules rather than from the cell's DNA. Just as one strand of DNA can be used to make its complementary strand, an RNA molecule holds the information needed to build DNA, and this can be done with a molecule called a reverse transcriptase. When cDNAs are made from RNAs, the introns have already been spliced out.

Building an artificial gene using a cDNA avoids some of the problems of producing human proteins in bacteria, but there may be others. As proteins are synthesized in the cell, they have to be folded into just the right shapes to become functional. After that, other molecules may trim off some of their parts. The molecules that do the folding and trimming may not exist in bacteria. Even if they could be added to the cell, such proteins probably would not be processed correctly.

Some molecules require other types of specialized processing—for example, they may need to be decorated with sugars. Some of these steps can be carried out only in plant or animal cells. So after perfecting techniques to create bacterial artificial chromosomes, scientists learned to insert circular artificial chromosomes in yeast cells (YACs) and mammalian cells (MACs).

Some of the earliest uses of molecular cloning included producing molecules needed by patients suffering from certain types of diseases. People with diabetes mellitus type 1, for example, are unable to produce the hormone insulin, and it cannot be extracted from other humans in sufficient quantities. Since the 1920s, researchers had obtained it from animals such as cows and pigs, but animal hormones were slightly different than those of humans and often provoked adverse reactions in the long term. After successfully engineering bacteria to produce foreign genes, Herbert Boyer began working on methods to use the cells to produce human molecules. His work attracted the interest of pioneering venture capitalist Robert Swanson

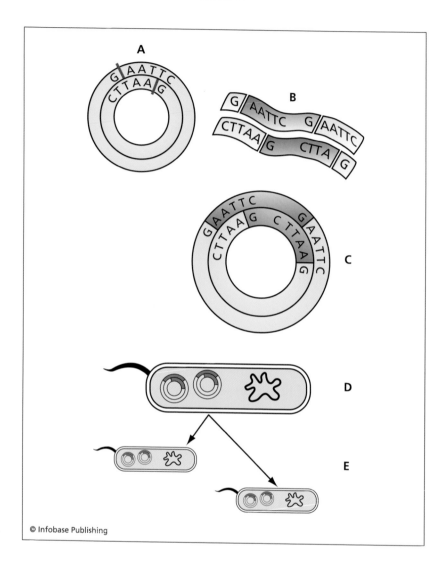

© Infobase Publishing

(1947–99), and the two men founded a company called Genen-tech with the goal of making insulin and other human proteins through genetic engineering.

DNA SEQUENCING

As the first steps toward modern genetic engineering were be-ing made, the British biochemist Fred Sanger (1918–   ) and his *(opposite page)* Inserting a foreign gene such as insulin into bacteria turns the cells into copying machines for molecules. A. Researchers begin with a circular bacterial chromosome called a plasmid and cut it at specific sequences (red lines) using a restriction enzyme. B. A version of the insulin gene is prepared with "sticky ends" matching the sequences of the cuts. C. Ligases glue the artificial gene into the plasmid. D. The plasmid is inserted into bacteria. E. When the bacteria reproduce, each new cell receives copies of the plasmid. Insulin hormone made in these cells is harvested, purified, and can now be used to treat people who suffer from diabetes.

colleagues in Cambridge were developing a new way to read DNA sequences. Sanger had already revolutionized the analysis of protein sequences. He discovered that proteins responded in different ways to electrical currents, and this could be used to obtain a readout of the amino acids making up a molecule. (Interestingly, the first protein he analyzed was insulin.) The work led to a Nobel Prize in chemistry for Sanger in 1958. His new strategy for sequencing DNA would eventually bloom into a successful, worldwide effort to obtain the complete genomes of humans and many other organisms and would lead to a second Nobel Prize in chemistry for Sanger in 1980. The first genome—the first complete sequence of an organism's DNA—was obtained in 1985. It was that of a bacteriophage called lambda, which consisted of 50,000 bases and took about five person-years to finish.

Sanger needed to obtain a readout of a DNA sequence that would identify the base in each particular location. The first step in this process was to *denature* a strand of DNA (which simply means to peel the two strands of the double helix apart using heat). One strand was kept and put into a solution with free-floating copies of the four bases and DNA polymerases. The polymerase used the pattern from the first strand to assemble free bases into a second strand. To start, it needed a primer, a chemical command telling it where to start work. One of Sanger's great innovations was to add another ingredient to the recipe: special versions of the four DNA bases, called *ddNTPs*. These are so similar to the natural bases that they are able to form base pairs with the strand. The difference, though, is that

Fred Sanger is one of very few people to win two Nobel Prizes: the first for new methods to sequence proteins; the second for revolutionary new methods of sequencing DNA that ushered in the era of high-throughput DNA sequencing and the human ge- nome project. *(Richard Summers, Wellcome Trust Sanger Institute)*

when a DNA polymerase adds a ddNTP version of a base onto a strand, it stops. One way to think of this is to imagine the polymerase laying new segments of track (normal bases) on a toy railroad. A ddNTP is like a piece of track with one broken end. The good end means that it can still be added to the last seg- ment, but no new pieces can be added at the end.

Sanger took the target strand that he wanted to sequence, copied it billions or trillions of times, and di- vided it into four batches in four test tubes. To each group he added four nor- mal bases and one type of ddNTP (one test tube, for example, contained the version for cytosine). Put- ting in low quantities of the ddNTP ensured that it would be chosen only rarely as the polymerase built the new DNA strand. (Using the railroad analogy again, it would be like providing the builder with lots of good track and a few broken segments.)

This happened in a random way—there was no way to pre- dict whether the very first cytosine, the very last, or a base in the middle would be exchanged for its ddNTP counterpart. In the end, each test tube contained many incomplete copies of the DNA molecule. Each began at the same place—the radioac- tive primer, where the polymerase started its work—but ended at the ddNTP (where it stopped). By measuring the length of the "tracks," Sanger could say, "30 bases from the beginning of the strand is a C." Since he carried out the same procedure in the other test tubes, with ddNTP versions of guanine and the other two bases, he could see that fragments that were 31 bases long ended with guanine, 32 bases meant thymine, position 33 was adenine, etc.

Sanger used a gel called polyacrylamide to measure the lengths of the fragments. This gluey substance is honeycombed by holes that are so tiny only molecules can pass through. He pressed the gel into flat sheets between panes of glass, mounted it vertically, and then poured in DNA molecules through separate channels (for the separately marked bases). Gravity pulled the fragments down, and the distance each one traveled was determined by its length. At a given time, he stopped the process and photographed it; the radioactive primer showed up as a dark stripe.

Most DNA sequencers today use fluorescent versions of ddNTPs that flash as they pass through multicolored lasers, in place of Sanger's radioactive primers. Additional improvements have increased the speed and accuracy of sequencing and allowed it to tackle much longer strands of DNA. Eventually, huge sequencing centers (such as the National Human Genome Research Institute, a division of the National Institutes of Health) were established in many countries around the world. Here scientists operate row upon row of sequencing machines that analyze DNA day and night. Coordinating their efforts, researchers across the world were able to analyze enormous amounts of DNA. By January 2007, researchers had completed the genomes of 1,250 viruses, 400 bacteria, and 46 higher organisms. That number is growing every day.

THE POLYMERASE CHAIN REACTION (PCR)

Most early types of molecular cloning used bacteria as factories to make genes, but the method did not always succeed in producing molecules that could do their jobs in the cell. Plus, at the end of the process, it was often difficult to extract the genes or proteins from the bacteria. The mid-1980s saw a solution: an ingenious new technique that permitted scientists to make millions or billions of copies of a DNA molecule in just a few hours, without using cells at all. The method required very small amounts of a sample to begin—the material from a single cell sufficed—and the DNA to be duplicated did not have to be fished out of the entire genome to begin. The method has had a huge impact on all types of research and medicine.

Its inventor, Kary Mullis (1944–), won a Nobel Prize in chemistry in 1993 for his work. In his acceptance speech, he

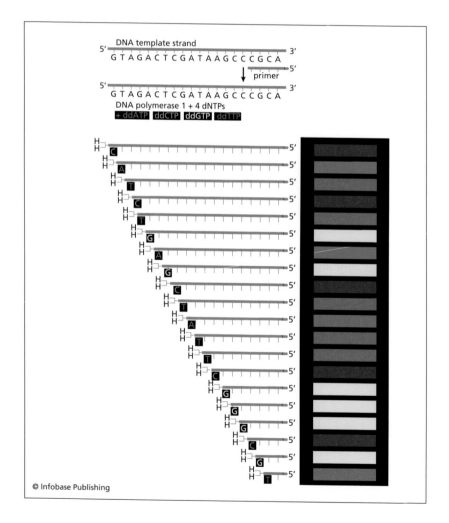

said that he conceived the technique while driving down a California highway. He was trying to figure out a way to analyze mutations in molecules; instead, he realized, he had discovered a method to make unlimited copies of any DNA molecule. At the time, he was working for Cetus, one of the world's first biotechnology companies. A dispute erupted over the ownership of PCR, whose applications in research and medicine would clearly make it extremely valuable. (Cetus gave Mullis a $10,000 bonus and later sold the patent to another company, Roche Molecular Systems, for $300 million.)

Mullis went on to take positions on political and social issues that gave him a reputation as an eccentric. For several years he denied that the HIV virus was the cause of AIDS (although PCR has provided evidence to prove it) and that humans have influenced global warming. He also claimed to have been abducted by aliens. In the 1990s, he was put on the list of witnesses to testify for the defense in the high-profile trial of O. J. Simpson, a football player and actor accused of murdering his wife, but was never called to the stand. Attorneys may have been worried about the jury's response to his unorthodox views.

PCR works by peeling two strands of DNA apart (through heating of the sample). Once they have separated, they are cooled again, allowing each strand to form a new partner by linking to free bases. A DNA polymerase sews the second strand together. The basic strategy had been tried before, but the polymerase that scientists used was destroyed by heat. This meant that the process had to be stopped and restarted between each round of copying. One of Mullis's contributions was to find a polymerase that could withstand the heat and

---

(opposite page) Copies of the DNA to be sequenced are divided into four test tubes. A molecule called a DNA polymerase starts to make a copy of the strand by picking up free bases and stringing them together. Each of the four batches contains a special version of one of the bases, called a ddNTP. At random times while copying, the DNA polymerase picks up a ddNTP instead of the normal base. This interrupts the process of copying, leaving molecules that are broken off at each position in the strand. The length of the strand and a radioactive label (or a fluorescent marker) show researchers which base is at each position.

carry out the reaction over and over again; the enzyme came from a bacterium that lives in an extremely hot environment.

To carry out PCR, one segment of DNA is preselected as the target to be copied. It is tagged at the beginning and end by short DNA sequences called initiators that tell the polymerase where to start copying. Once a round of copying is completed, the process is repeated. The new double strands are separated by heating and then are cooled again so that the next round of copying begins. It only takes a few hours to carry out 30 or 40 cycles, each of which doubles the amount of DNA, soon producing huge quantities of molecules.

PCR has become a standard tool in molecular biology, medical research, and forensic science. Because very little material is required to begin, it can be used to amplify tiny amounts of samples from patients or traces of DNA left at crime scenes. It has been used to extract and amplify damaged DNA from fossils such as the bones of Neanderthals and mammoths, to search for traces of tuberculosis and other diseases in mummies, and to solve historical mysteries.

The method can also be used to trace the evolution of viruses and other disease organisms, that is, to determine where they came from and how they are transmitted through a population. The HIV virus undergoes such rapid mutations that there are small differences between the viruses found in each person who catches the disease. These individual differences are carried along as the virus goes on to infect a new victim, where it also acquires unique features. This permits scientists to determine the path of an infection—who has given the virus to whom. Without a rapid and efficient method like PCR, it would be impossible (or very impractical) to carry out such studies.

(opposite page) PCR revolutionized molecular biology by giving researchers a simple method to make huge numbers of copies of a DNA sequence they wished to investigate. A. Heat is used to separate the two strands of a DNA molecule. B. Primers are added to the beginning and end of the target sequence. C. A DNA polymerase is added; it builds a partner for each of the two strands. D. The process is repeated over and over until enough copies have been made.

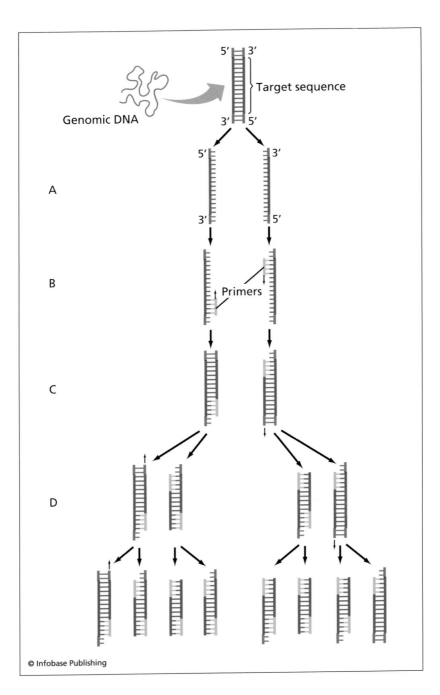

MAKING AND MARKETING GENETICALLY MODIFIED PLANTS AND ANIMALS FOR FOOD

Genetic principles have been used to manipulate domestic plants and animals ever since the rediscovery of Mendel's laws at the beginning of the 20th century. Farmers and researchers hoped that the new science could increase the productivity and nutritional value of crops and animals, partly motivated by concerns about whether there would be enough food for the rapidly growing populations in industrialized nations. Scientific breeding methods improved corn, tomatoes, and many other crops. But improvements depended on natural mutations that gave plants desirable properties. Scientists could not plan or predict what these might be.

Today the same kinds of concerns—and other motivations—have prompted scientists to use the new tools of genetic engineering to modify crops and other organisms. Food production cannot keep up with population growth. At the World Food Summit in Rome in 1996, experts stated that the world would have to double its production within the next 30 years just to keep pace. It has been estimated that 800 million people on the planet currently suffer from malnutrition and starvation. That number will rise tremendously unless more food can be produced, and genetic engineering has been seen as one means to that end.

The idea was to create *genetically modified organisms* (GMOs) by directly modifying the genes of crops. As well as improving their size, taste, shelf life, or nutritional value, they could be protected from insects, fungi, and other parasites. Chemical pesticides had reduced these problems but caused others: Over the long term, the substances that killed parasites entered the soil and the food chain. This could damage the DNA of animals and people that ate the plants, causing cancer, allergies, and other health problems. On one hand, genetics might be able to ward off pests without such dangerous side effects. On the other, the members of growing ecological and environmental movements protested that genetic engineering might upset delicate balances in nature.

Companies interested in creating new foods—for humanitarian and economic reasons—claimed that farming had always produced highly artificial crops and they began to investigate ways to improve foods through genetic engineering. In 1994, a company called Calgene brought the first crop developed with genetic engineering onto the market: the "Flavr Savr" tomato. Scientists had discovered that a protein called polygalacturonase played an important role in how tomatoes rot because it softens cell walls as the fruit ripens. By inserting a second gene that interefered with the protein, the tomatoes could be stored longer without losing their taste.

The U.S. Food and Drug Administration (FDA) examined the plant, deciding that it did not pose a health hazard to people and could be put on the market without special labeling. Although the tomato was more expensive than other brands, customers in the United States were enthusiastic. In Europe, tomato paste made from the Flavr Savr strain was brought onto the market at a very low price as an experiment to see if consumers would buy GM products. But Calgene had not used the best strain of tomato to begin with, and the company had little experience in growing and marketing foods. So in the long term, Flavr Savr lost out to other long-lasting, non-GM brands that customers preferred. Even so, the following year Calgene was bought by the company Monsanto, which has become a major producer of many types of genetically modified foods.

In Europe, the public acceptance of GMOs quickly plummeted as consumers became concerned that they might be eating such foods without knowing it and that there might be unknown risks. Protestors demanded strict governmental controls (such as bans on imports, or at least clear labels marking food as a product of genetic engineering). The change in attitude was partly due to outbreaks of a deadly disease called *bovine spongiform encephalopathy,* or mad cow disease, which animals caught by eating infected food. The disease was then passed along to humans who ate the animals. While many Europeans felt that their governments were no longer doing a good job of regulating foods, most Americans have remained more receptive. This has caused stress in international trade as U.S. companies have found it hard to market GM products in Europe.

GM tomatoes were quickly followed by soybeans, cotton, and maize. Some of the new varieties improved the nutritional value of staple foods that are the core of people's diets in many parts of the world. Corn and golden rice lack vitamin A, which is essential to the development of the eye; genes have now been added to provide it. The changes have helped reduce blindness and other symptoms of malnutrition that have plagued children throughout the world. Plants have also been made resistant to the herbicides used to kill weeds. Tomatoes, cotton, corn, and many other crops fall prey to caterpillars; researchers have added a natural toxin, a protein from the bacterium *Bacillus thurinienses,* that kills the insects. Sweet potatoes in Africa have been made immune to viruses. Changes in species of rice have produced strains that can survive floods, and other plants have been modified to tolerate high levels of salts or acids in the soil.

Like the natural breeding practices described in chapter 1, genetic engineering has also been used to make plants more aesthetic or useful in other ways. One common technique has been to add a gene called monellin, taken from the seeds of an African plant, to tomatoes and other vegetables. This molecule is a very powerful natural sweetener; even in small quantities, it enhances flavor. And the efforts have not been uniquely aimed at altering or improving foods. Scientists discovered a gene in a species of bacterium called *Alicaligenus eutrophus* that produces a type of plastic called polyhydroxybutyrate. By inserting this gene into cotton plants, they hope to create new types of cloth. They also hope to put the techniques to use to create new medicines, inserting genes into bananas, potatoes, or other foods. Studies are currently being carried out to discover whether this might be an effective strategy to cure diabetes or even to vaccinate people against the flu or other viruses.

All of these modifications require delivering foreign genes into plants. In 1986, Cornell University professor John Sanford found a new method to do so—inspired by the BB gun he used to chase squirrels from his garden. With the help of colleague

Edward Wolf, he built a miniaturized gun that could shoot genes into plants. Another technique involved a bacterium commonly found in the soil, called *Agrobacterium tumefaciens,* known to infect potatoes and other plants. In doing so, it deposits its own genes in the plant's cells. This and other strains of bacteria, viruses, and yeasts have been developed to "transfect" plants with monellin and other genes.

The number of GM crops continues to increase dramatically, particularly in the United States, Argentina, Canada, and China. Recently, it has been estimated that about 75 percent of foods on the shelves of stores in the United States contain at least one GM ingredient. In other countries, the trend has grown at a somewhat slower pace, but overall GM crops are winning an increasing share of the world food market. By 2005, approximately 60 percent of the world's soybean fields, 28 percent of the cotton, and 14 percent of the maize were devoted to GM crops.

Decisions to develop and grow GM foods are based on the profit they are expected to bring as well as other motives. Businesses have sometimes engaged in questionable practices to gain an advantage over their competitors to the detriment of farmers and economies in developing nations. The practices have also raised new legal issues such as questions of ownership. The creation of a new crop requires a huge investment in basic research, laboratory experiments, costs of growing, and risk assessments. Companies need to recapture these costs through profits, which are best ensured by maintaining ownership of their crops. The idea that a species can be owned is a complex ethical problem that is discussed in the next chapter.

There has been a growing interest in the production of GM animals for foods as well, but the efforts have met with technical, ethical, and legal challenges. It is much more difficult to develop GM animals than plants. Often a new plant can be grown from an existing one simply by taking a single cell. In animals, new genetic material must be introduced into the very early embryo, so that the animal's egg or sperm cells contain the gene. The methods are not perfect, and many generations may be needed to obtain a strain with the gene.

Getting approval to engineer animals for food has been more difficult than growing and marketing GM plants. The first such animal food to go on the market may be a fish called the super salmon, which grows to full size much quicker than its natural relatives. In the wild, salmon grow quickly during warm weather but hardly at all in cold seasons. Adding an "antifreeze" gene resulted in fish that grew to seven or eight pounds in 18 months, about twice the normal rate, and more work has led to salmon that grow even faster. The FDA has been conducting lengthy tests on the salmon to ensure that eating it poses no health hazards.

In 2000, when the issue arose, the FDA was the only governmental agency with the legal power to make the decision. Critics remarked that the FDA would probably do a good job of determining the safety of the fish as a food, but this is not the only concern. The Environmental Protection Agency and U.S. Department of Agriculture would undoubtedly be better candidates to estimate its environmental impact. If the strain were to escape farms and enter the wild, it would mate with native fish. Scientists could modify the fish so that it was sterile. However, during its lifetime, it would still compete with native fish and probably disrupt the ecologies around them.

Other efforts are under way to create pigs that produce leaner meat and to use animals as factories for drugs such as insulin, described earlier in the chapter. The same strategy was used to make another hormone called erythropoietin, which stimulates the development of red blood cells and has been used as a treatment in anemia and some forms of kidney disease.

Another use of genetic engineering is to create drugs that might be delivered to people through animal foods: chickens whose eggs contained antibodies, or bananas containing vaccines. Animals have been engineered to produce milk that contains human proteins—usually not for drinking, but because it is fairly easy to extract the molecules from the milk. In 1988, Ian Wilmut and his colleagues at the Roslin Institute in Scotland created a sheep that produced a human protein called alpha 1-antitrypsin, used to treat emphysema. (Wilmut later became famous when his group cloned the first mammal, a sheep named Dolly, discussed in chapter 6.) Pigs have been engineered to produce milk that contains Protein C, an agent in human blood that helps it to clot when there is an injury. The milk contains hundreds of times the concentration of the molecule that is normally found in human blood.

KNOCKOUTS, KNOCK INS, AND OTHER METHODS TO STUDY GENE FUNCTIONS

Genetic engineering has given scientists several new methods to investigate the roles that genes play in the lives of cells and organisms. Some of the techniques include the following.

- knockouts, which delete a gene
- knock ins, which add a gene to a cell or organism that does not normally have it
- overexpression studies, which raise the amount of RNA and/or proteins produced from a given gene

The first genetically modified animals were made by Hermann Muller using radiation that introduced random changes in DNA bases. Later, scientists began to use a chemical called ENU, for N-ethyl-N-nitrosurea, which changes single base pairs in DNA. These and other types of mutagens often led to knockouts because they disturbed a gene sequence and led to a defective protein or made it impossible for cell to produce a certain molecule at all. But sometimes the experiments had other effects. Mutagens can affect any region of DNA, including sequences that are responsible for controlling whether a gene is switched on or off or when and where it becomes active in an organism. A mutation that prevents a gene from being activated causes a knockout. But if the change instead makes it impossible to turn the gene off the result is constitutional activation. This can lead to disaster. Suppose, for example, that the function of a gene is to tell a cell to divide. During most of the cell's lifetime, this gene should be silent. But if it becomes permanently active, the cell may spin out of control, dividing over and over and leading to tumors.

Beatrice Mintz: A Pioneer of Mouse Genetics

Beatrice Mintz (1921–) was a young biology professor at the University of Chicago in 1953 when James Watson and Francis Crick solved the structure of DNA. Just a little more than a decade later, after moving to the Institute for Cancer Research in Philadelphia, she began a series of pioneering experiments with mice embryos that have played a vital role in modern disease research. Another decade after that, she had established a reputation as a world-renowned mouse geneticist.

The first successful genetic experiments required a deep understanding of bacteria and the structures and functions of their genes. Learning to manipulate the genes of mice required a profound understanding of the earliest stages of the animal's development. Bacteria often absorbed foreign genes directly from the environment, but the cells of mice and other mammals had defenses that prevented this from happening. How could genes from other species—or artificial molecules built in the laboratory—be transplanted into mouse cells? Mintz's work was crucial to finding an answer.

In the early 1960s, Mintz found a way to combine embryonic cells of different mice into a single animal. She took several early embryos and squeezed them together. They fused to form one new embryo. When she implanted it into a mother, it developed into a complete, healthy mouse, even though it was a chimera—a composite of several different genomes. Later the animal could be analyzed to discover which cells had produced different parts of the body. This was useful in medical research, as it would allow scientists to follow a genetic disease back to its origins in the very early embryo.

In the 1970s, Mintz collaborated with the German scientist Rudolf Jaenisch (1942–), now at the Whitehead Institute for Biomedical Research in Massachussetts, to show that foreign DNA could pass into the hereditary material of mice. The two researchers infected embryos with the *Simian virus 40* (SV40). It is a retrovirus, like HIV, which means that it reproduces by writing information from its genes into the cells that it infects. Retroviruses usually target particular types of cells—HIV, for example, infects certain types of white blood cells. Mintz and Jaenisch proved that SV40 added genetic material to the mouse germline—egg and sperm cells—and that the new genes were then passed along to the mice's offspring. If viruses could introduce new genes into the mouse genome, then maybe the same thing could be done with genes that had been modified by genetic engineers.

At the same time, Mintz was continuing her work with mouse chimeras, trying to understand the development of cancer. She discovered that when cells from embryos of mice prone to cancer were combined with those of healthy mice, the embryos would grow into healthy animals that lost their susceptibility to cancer. This said something about the way cancer developed, and it has been a key theme of her research ever since.

In 1980 and 1981, Jon Gordon and Frank Ruddle developed a technique to inject foreign DNA into the nuclei of mouse egg cells using a microscopic pipette. In some cases, this led to the integration of the gene into the genome as the cell divided to form an embryo. It also appeared in the DNA of the new mouse's reproductive cells, giving scientists a way to create *transgenic* mice. Within a year, six groups, including that of Mintz, had used the technique to produce a variety of mice with alterations in several genes.

Since the 1980s, Mintz and her colleagues have produced a range of different transgenic mice, chiefly to study how cancer arises and develops into life-threatening metastases. She has also worked to develop *gene therapies*, hoping to train immune system cells to recognize and respond to molecules on the surfaces of cancer cells.

When scientists applied mutagens to early embryos such as fly eggs, the damage was often inherited by every cell in the animal's body as it grew. If the animal was further along in its development, the damage might be restricted to specific organs or particular types of cells. This is usually what happens in cancers caused by radiation or chemicals. Substances in cigarette smoke, for example, cause frequent mutations in lung cells. Sometimes the changes are harmless or kill the cells outright. But if they affect genes that control when cells divide or how they develop, the cells may become cancerous.

While X-rays, ENU, and other substances cause high numbers of mutations, their effects are random. The study of animals produced in this way is called forward genetics: looking at animals for disease symptoms or other features and then tracking down the genes that have been changed. Obviously, this is a slow process, and a researcher interested in a particular disease may have to wait a long time to find an animal with the symptoms he or she is interested in—if such an animal ever appears at all.

The successes of genetic engineering suggested that it might be possible to take the opposite approach (reverse genetics): to make precise, targeted changes and watch what happened to the animals. This would give scientists a way to study how defective molecules affected cells, embryos, and adult organisms and to try out new types of therapies. But introducing mutations in animals turned out to be far more complex than engineering bacteria.

TRANSGENIC ANIMALS AND MODELS OF HUMAN DISEASE

One of the main reasons to create transgenic animals is to learn about human diseases and even to try out potential treatments. The close evolutionary relationship between humans and animals such as the mouse means that the species have related genes that often function in similar ways. Defects in genes often cause similar problems for the two species, so studying mice with mutations allows researchers to probe diseases in ways that would be very difficult or unethical in humans. Not all of the model animals are products of genetic engineering. Some were developed with the help of mutagens; others were discovered through experiments that inbred mice or other organisms over many generations and then screened the offspring to find animals with symptoms similar to human diseases.

Genetic diseases result from errors in DNA. These may be inherited from a parent, or they may develop spontaneously because of mutations or damage to a person's DNA. A change in one or more letters of the genetic code makes cells unable to produce a necessary protein, or they produce a faulty version of the molecule. In some cases, the protein itself is properly formed but is produced at the wrong time or place in the body.

Several thousand diseases are known to result from errors in single genes; others are the result of combinations of errors. Many disease-causing mutations are dominant, which means that inheriting one copy of the gene causes health problems. If the gene is recessive, it is most dangerous when a person inherits a mutation in the same gene from both parents. Inheriting a single copy may lead to milder symptoms, or they may not occur at all.

In 1984, Philip Leder and Timothy Stewart used genetic engineering methods to insert a cancer-causing gene called an *oncogene* into a strain of mouse in order to use the animal for cancer research. When they applied for a patent on this strain, a new ethical issue was raised: Could individuals and companies "own the rights" to living organisms? The U.S. Supreme Court had just ruled that they could, and the U.S. Patent Office approved the application.

In the meantime, much more refined methods have been developed that allow researchers to shut down a gene only in specific tissues at specific times. This is useful, because genes often have multiple functions. Removing a gene entirely may kill an organism in an early stage of its life or cause other severe problems that prevent scientists from studying another role that the molecule plays later in a specific tissue.

An early knock-in technique in animals involved injecting foreign genes into the nuclei of fertilized eggs that were grown in a test tube and then implanted into a mother. A limitation of this method was that there was no way to control where the gene landed in a genome, and genes' positions often influence their activity. The problem was overcome when geneticists Martin Evans at Cambridge University, Oliver Smithies at the University of Madison–Wisconsin, and Mario Capecchi of the University of Utah developed methods to control where genes landed in animal cells by replacing an existing gene with a defective or altered version. The first step in this process, which they called homologous recombination, was to obtain animal embryonic stem cells—the first cells produced when a fertilized egg divides. Then they injected an artificial gene, hoping that it would replace the original. Like genetic engineering techniques described earlier in this chapter, this was possible because the cells contained enzymes that cut and mended specific DNA sequences. The artificial gene had DNA sequences at each end that were identical to the sequences flanking an existing gene. The scientists hoped that this would fool the cell into thinking that the loose gene had been cut out by mistake and needed to be put back in its right place, replacing the original. Adding probes to the artificial genes allowed them to detect whether this had happened and whether the gene had landed in the right place. The doctored cells were put back into embryos, some of which survived and produced the new version of the gene in at least parts of their bodies.

By the mid-1990s, scientists across the world were routinely using genetic engineering to study gene functions, to make animal models of diseases, and to create new types of plants and animals. But many of the methods were still quite crude and did not give researchers the level of control they needed to understand how genes affected organisms. That situation has been changing rapidly with the completion of genome sequences and the development of new types of biotechnology; these themes are addressed in the next chapter.

Today scientists have produced hundreds of mouse models of human diseases, using these techniques or more sophisticated *"conditional knockout"* strategies discussed in the next chapter. The following table lists a few mutations in mice genes and the human diseases that are being studied in connection with them.

EXAMPLES OF MOUSE MODELS OF HUMAN DISEASES

Mouse Gene	Human Disease	Common Symptoms in Humans
Endothelin 3	Waardenburg syndrome	Moderate to severe deafness; patches of white pigment on the skin
Myosin VIIa	Usher's syndrome type IB	Deafness and problems with balance caused by changes in inner ear structures
Rhodopsin	Retinitis pigmentosa-4	Blindness in middle age due to degeneration of photoreceptor cells
Procollagen type 1	Osteogenesis imperfecta	Development of weak or brittle bones
Fibrillin	Marfan's syndrome	Long limbs and fingers; defects of the heart valve and aorta
Apolipoprotein E or Amyloid beta precursor protein	Alzheimer's disease	Death of neurons, leading to memory loss, loss of language and cognitive abilities, and dementia
Dystrophia	Muscular dystrophy	Progressive weakness of skeletal muscle tissue and defects in muscle proteins
Huntington disease homolog	Huntington's disease	Progressive death of neurons in specific regions of the brain, loss of control of movement, behavioral and cognitive defects
Breast cancer 1	Breast cancer	Breast cancer
MutL	Familial colon cancer	Colon cancer

(Table continues)

EXAMPLES OF MOUSE MODELS OF HUMAN DISEASES (*continued*)		
Mouse Gene	Human Disease	Common Symptoms in Humans
Neurofibromatosis type 1	Neurofibromatosis type 1	Tumors that form around nerve cells
Coagulation factor VIII	Hemophilia A	Inability to stop bleeding after injuries
Hemoglobin beta gene cluster	Sickle cell anemia	Improperly formed red blood cells
Cystic fibrosis transmembrane conductance regulator homolog	Cystic fibrosis	Thick mucous production connected to lung infections and a progressive failure of organs such as the lungs, liver, and pancreas
Trp53 tumor suppressor gene	Cancer	Susceptibility to a wide range of cancers

A large number of diseases have been linked to defects in multiple genes. It is usually very difficult to pin down the molecules that are responsible and create animals with multiple knock-outs. Some progress has been made through studies of inbred mice and rats; inbreeding often leads to populations of animals with combinations of disease traits.

DNA FINGERPRINTING

The revolution in biotechnologies such as DNA sequencing in the 1980s led to some unexpected applications, such as *DNA fingerprinting*. Sequences could be used to determine the person who had left DNA behind at a crime scene or to determine relationships between people. The method was invented in the early 1980s by a young scientist named Alec Jeffreys, who was doing research into the evolution of genes at the University of

Leicester in Great Britian. Jeffreys, his colleague Polly Weller, and other members of the lab had obtained the sequence of a gene called myoglobin from many different species—mice, baboons, and even the tobacco plant—intending to see how macroevolution had changed its sequence over vast stretches of time. They were equally interested in microevolution in humans—the much smaller genetic changes that occur within a single species.

Myoglobin had an interesting characteristic: It contained repeated sequences of DNA that changed in just a few generations—much more quickly than the rest of the genome. Jeffreys immediately realized that such unusual bits of code (called *minisatellites*) could be used to obtain genetic fingerprints of people. By chance, the team had collected DNA samples from one of the lab technicians and his two parents. The pattern clearly showed that the technician had inherited some of the genetic markers from his mother and others from his father. Those were different than the markers of other members of the laboratory.

When the team published its scientific paper, the story was picked up in the popular press, where it was noticed by a lawyer who had been working on a legal dispute regarding a family from Ghana. When one of the children entered the country, authorities suspected he might not be a real member of the family. The case was complicated by the fact that the father was missing. Jeffreys's laboratories ran tests that examined the mother and several of the children. This allowed them to reconstruct a profile of the genetic markers of the missing father and prove that the boy was indeed his child. The landmark case began a long tradition of using DNA fingerprinting to answer questions about paternity and to match samples from crime scenes to suspects. The method has identified murderers and other criminals in thousands of cases by matching DNA to suspects; it has also cleared people formerly convicted of crimes by showing that their DNA did not match samples taken from crime scenes.

Another interesting application has been to clear up several historical mysteries. For example, it proved that Thomas Jefferson had seven children by his slave Sally Hemings, the African-American relative of his deceased wife. More recently, the method has shown that the entire family of Czsar Nicholas Romanov II was murdered and buried on the eve of the Russian Revolution—contradicting rumors that his daughter Anastasia might have survived. In 1920, a woman named Anna Anderson claimed to be the missing girl, setting off a worldwide controversy about the fate of the Romanovs' daughter. However, a young woman from Berlin claimed to recognize Anderson as a former roommate and the daughter of a Polish farmer. An analysis of Anderson's DNA, carried out at Pennsylvania State University in 1995, confirmed her relationship with the Poles and ruled out one to the Romanovs. In the meantime, remains of all the family members have been found, and researchers are sure that Anastasia was murdered along with the rest of her family.

5

Genetic Engineering in the Age of Genomes

The discovery of how information contained in DNA becomes transformed into RNAs and proteins marked the beginning of molecular biology and permitted the development of genetic engineering. Within a few decades, scientists learned to remove, transplant, and alter genes and began to manipulate organisms by changing the hereditary information contained in their cells. They even managed to simulate human diseases in animals by engineering the molecules known to be defective in genetic diseases. Yet despite such advances, made in a spectacularly short time, science is still unable to cure most human ills and lacks a deep understanding of what happens within organisms. One reason is the incredible complexity of what happens in cells, where single events are often the result of interactions between hundreds or thousands of different types of molecules. Even today's most advanced computers are unable to analyze or simulate these events. These are some of the challenges for the future, and this chapter shows how current work in genetics is beginning to address them.

THE COMPLEXITY OF GENOMES

Watson and Crick's dogma, DNA makes RNA makes proteins, is still a good basic outline of the first few steps in how organisms use the information contained in their genes. But the last 30 years have revealed that the route between genes and proteins is a labyrinth of forking paths rather than a straight road. Each step along the way is crucial to understanding how organisms develop, how they evolve, and how they respond during diseases and therapies.

The complexity of this pathway begins with the genome. Until about two decades ago, scientists' knowledge of DNA was incomplete; they had sequenced only small regions of the genomes of humans and other organisms, and it was even impossible to estimate how many genes a species possessed. Then in the 1980s, advances in DNA sequencing suggested that it might be possible to read the entire human genome. The U.S. Department of Energy conducted a series of workshops to determine whether it could realistically be done, how much effort would be required, and how much it would cost. In 1990, the project was formally launched by the U.S. Department of Energy and the U.S. National Institutes of Health. Researchers estimated that it would probably take 15 years of intensive work by laboratories across the world to obtain a complete sequence of human DNA, taking into account advances in DNA sequencing technology that were sure to happen. James Watson was appointed to head the project. He was replaced in 1993 by the physician and geneticist Francis Collins (1950–   ).

Researchers all over the world made unofficial bets on how many genes the project would uncover. Guesses ranged from about 30,000 to 150,000 genes. In the end, the answer lay at the lower end of the scale. Although the analysis of the genome is not yet complete (it was finished in 2003, but the data is still being interpreted), most scientists now believe that it contains fewer than 30,000 genes; in fact, the number is probably closer to 20,000. This seems amazing, because organisms considered much simpler than humans have nearly as many: The tiny fruit fly has about 13,000 genes, and a tiny worm called *C. elegans* has approximately 18,000. On the other hand, rice, maize, and many other plants probably have far more genes than humans.

Even "small" numbers can create complex organisms, because there is a vast number of recipes by which genes can work together to produce results. At any one time, a typical human cell uses only about 20 percent of its genes to make proteins. Each type of cell uses a different set, and the pattern changes as various things happen to the cell. This pattern has many effects. For example, it changes stem cells into sprawling, treelike nerve cells, or doughnut-shaped blood cells; it helps them take up their proper positions in the body and tie themselves tightly to their neighbors; and it sets the clock of cell division. Some genes become active at precise moments in order to kill off cells that have outlived their functions. Others are switched on only during an infection.

Matters get more complicated because genes make up only a tiny proportion of the total human DNA sequence. The genome reveals that the vast majority—probably at least 98.5 percent—does not encode proteins. Originally, this appeared to be "junk," an accumulation of meaningless fragments of DNA that have copied themselves over and over or ancient genes that have lost their functions through mutations. In the meantime, a lot of the junk has turned out to have functions after all. These are discussed in later sections of this chapter.

INTRONS AND ALTERNATIVE SPLICING

Chapter 3 introduced the fact that noncoding DNA sequences, called introns, often appear right in the middle of genes. This discovery, made in 1977, led to the 1993 Nobel Prize in physiology or medicine for British biochemist Richard Roberts, working at Cold Spring Harbor Laboratories in New York, and the American geneticist Phillip Sharp of MIT. But only with the completion of the human genome have scientists seen how extensive splicing truly is. They have also learned a great deal about the functions of this process in cells.

An average human gene holds 8.6 introns, but some are far more complex. The gene that encodes the giant muscle protein called titin contains 362 introns. The current record holder is a gene in the fruit fly called Dscam. It can potentially generate 38,016 different proteins. Each of these might have its own specialized functions. Just as trains can be assembled with made-to-order contents for different customers, exons as well as introns are sometimes left out of RNAs, making it possible to obtain hundreds or thousands of unique RNAs (and thus different proteins) by combining different subunits of a single gene.

Introns in a single human gene frequently total tens of thousands of nucleotides and are, on average, five times the length of the exons. This may have an impact on the evolution of human genes, as revealed in a 2002 study by the evolutionary biologist Cristian Castillo-Davis of Harvard University, working with Eugene Koonin and Fyodor Kondrashov of the National Center for Biotechnology Information in Maryland. Transcribing RNA is a slow and energy-expensive process: Making RNAs from a single gene with huge introns can require several minutes and thousands of energy molecules called ATP. The authors found that the introns of frequently used genes are, on average, 14 times shorter than those of rarely used genes. They conclude that natural selection has been shortening introns in the most common genes, saving time and energy.

Once the cellular machinery needed to carry out alternative splicing evolved, it could be put to use in many different ways. Mixing and matching modules produces proteins that behave differently. They help to create diverse types of cells and figure prominently in the development of different tissues. Alternative splicing of an RNA called Slo, in the ear of the chicken, improves the bird's hearing by giving it cells sensitive to different frequencies of sound. In flies, three critical proteins are spliced differently in males and females and play a crucial role in causing their bodies to develop differently. Although females have two X chromosomes and males only one, females do not produce twice the amount of proteins from the genes on the chromosome, thanks to the differences in these proteins. In 2005, Diane Lipscombe and her colleagues at Brown University in Rhode Island found that alternative splicing is especially common in the brains of mice and other mammals. Spliced forms of particular RNAs are crucial to memory and learning.

Splicing is likewise an important factor in a wide range of diseases. Half of the people who suffer from neurofibromatosis, a severe genetic disease in which tumors develop alongside nerves and other tissues, have mutations that change the splicing of RNAs made from the neurofibromin gene. Patients with beta thalassaemia suffer from anemia because of incorrect splicing of the beta-globin gene. Other examples are changes to the BRCA1 gene (linked to breast cancer) and the CFTR gene (leading to cystic fibrosis). It is estimated that about 50% of the mutations in exons that cause disease affect the way RNAs are spliced. Tumors and neurodegenerative diseases are often accompanied by unusually spliced RNAs that are not normally found in healthy tissues. But the cell has defense mechanisms to protect it: In some cases, it can identify improperly spliced RNAs and break them down before they do harm (see sidebar).

NONCODING RNAS

As scientists began to investigate the genome in detail, they were surprised to find that the cell produces a large number of RNAs that are transcribed from DNA but do not encode proteins. Why would the cell spend so much time and energy on "useless" molecules? The phenomenon was a complete mystery until researchers found out that they had important functions—many of them caused the destruction of messenger RNAs. In doing so, they blocked the synthesis of proteins.

The discovery was made in an unlikely way. In the late 1980s, scientists in a biotech company in California were using genetic engineering to try to alter the color of petunias. Geneticist Richard Jorgensen's laboratory inserted an extra copy of the gene responsible for purple pigment, expecting that the extra gene would cause the production of more pigment protein and a more intense color. Instead, the result was a completely white flower without any pigment at all. A look into the cells showed that they were using both genes to produce RNA molecules, as expected, but that the RNAs were somehow interfering with each other. Laboratories across the world were discovering the same phenomenon in other plants, and soon it would be observed in animals as well. Finding out why this happened took several more years.

Quality Control: How Cells Detect Defective Genes

In 1979, Regine Losson and François Lacroute of the National Scientific Research Center in Strasbourg, France, discovered that the cell has a system to inspect RNAs for "quality control." Nearly three decades of research have shown that the system is not perfect, but it manages to protect eukaryotic cells from the dangerous effects of most mutations. Defects in genes that change the shape, structure, or functions of a protein usually have bad effects on the cell. Mistakes in splicing can also produce such molecules, so the inspection system—called *nonsense-mediated mRNA decay* (NMD)—needs to be on alert all the time.

NMD and splicing are closely connected. Cells remember that an RNA has been spliced by attaching a cluster of proteins to the sites where introns have been removed. The cluster consists of at least six proteins and is called the *exon junction complex* (EJC). Later, if the EJC seems to be in an inappropriate place, the RNA is destroyed by NMD.

Understanding how this works requires a closer look at how RNAs are translated into proteins. This job is carried out by the ribosome, a machine made of several parts. It docks onto a messenger RNA, reads its code, and assembles a chain of amino acids that matches the sequence. Any EJCs on the molecule are simply moved out of the way. At the end of the protein-encoding part of the RNA, the ribosome encounters a stop codon, which signals the end of the coding region and releases the finished protein. Mutations often alter the spelling of an RNA so that a stop codon appears somewhere in the middle of the molecule. This creates a code within the RNA that does not make sense to the cell. It looks like nonsense because of its position relative to an EJC.

All RNAs have a stop codon, but it should come after the exons and introns. This means that normally an RNA

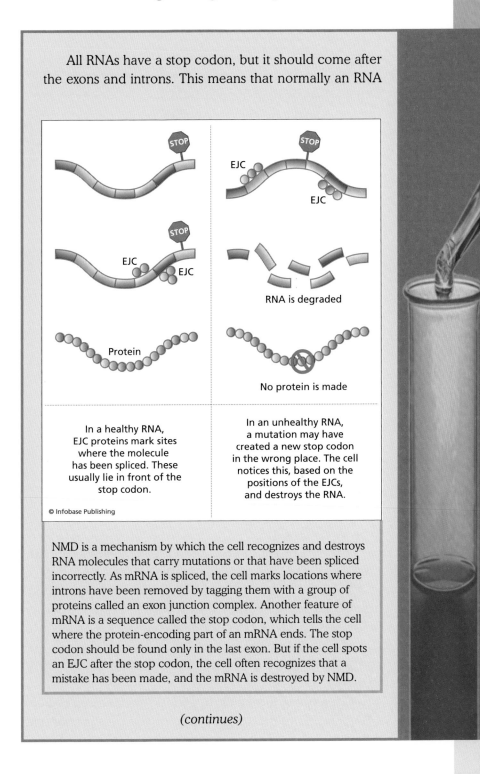

In a healthy RNA, EJC proteins mark sites where the molecule has been spliced. These usually lie in front of the stop codon.

In an unhealthy RNA, a mutation may have created a new stop codon in the wrong place. The cell notices this, based on the positions of the EJCs, and destroys the RNA.

© Infobase Publishing

NMD is a mechanism by which the cell recognizes and destroys RNA molecules that carry mutations or that have been spliced incorrectly. As mRNA is spliced, the cell marks locations where introns have been removed by tagging them with a group of proteins called an exon junction complex. Another feature of mRNA is a sequence called the stop codon, which tells the cell where the protein-encoding part of an mRNA ends. The stop codon should be found only in the last exon. But if the cell spots an EJC after the stop codon, the cell often recognizes that a mistake has been made, and the mRNA is destroyed by NMD.

(continues)

(continued)

is not spliced behind the stop codon—at least not very far behind it—and no EJC is located there. Recognizing that this has happened is the key to NMD. If an EJC comes more than about 50 nucleotides after a stop codon, the process of translation is interrupted, and other molecules carry the RNA away and break it down.

But the system is not perfect. Some RNAs escape NMD and produce harmful proteins. Even when NMD works, the result may be disease, because the process may destroy an RNA that is damaged but nevertheless necessary. In 1989, Lynne Maquat's laboratory at the Roswell Park Memorial Institute in New York showed that NMD contributes to beta thalassaemia, the most common genetic disease in the Western world. Beta thalassaemia reduces the body's production of *hemoglobin,* which is needed to carry oxygen through the blood. The disease arises in people who inherit a mutant form of a gene called beta globin. NMD catches the mutation, and the body breaks down beta-globin RNA—removing an important molecule. In this case, what is normally a safety mechanism ends up attacking the body.

DNA always forms a double strand. RNA usually remains single stranded, even though it is made of the same type of building blocks, because cells usually do not produce RNAs with complementary sequences that would bind to each other. If that happens, however, the double-stranded RNAs are usually attacked and destroyed by cellular defenses. The cell normally interprets such molecules as alien, because many viruses contain them.

The first examples turned up in genetic engineering projects such as Jorgensen's, but later this type of *RNA interference* was

Until recently, NMD was considered to be little more than a means of trapping RNAs that contained errors; now it is known to be a more general tool that the cell uses to control the quantity and quality of certain molecules. This happens because the normal process of alternative splicing sometimes produces RNAs with nonsense codons: For some reason, the cut-and-paste operation produces a bit of nonsense code in the middle of an RNA. In 2004, R. Tyler Hillman, Richard Green, and Steven Brenner of the University of California, Berkeley, carried out a computer analysis that showed that about one third of the time, alternative splicing places a stop codon more than 50 nucleotides in front of a splice site. This activates NMD, which eliminates most of the RNA before it can be transformed into proteins.

The same year, Harry Dietz's group at John Hopkins University School of Medicine in Maryland studied this effect in the cells of mammals. They completely shut down the NMD machinery by removing a protein called Upf1, which is essential for the process. This changed the behavior of a huge number of genes: About 10% of the genes they studied became more productive, probably because spliced forms were slipping through that normally would have been caught by NMD and destroyed.

found to happen naturally in cells. Animal and plant genomes contain the recipes for a huge number of noncoding RNAs, usually very small ones called *microRNAs,* whose sole purpose seems to be to lock on to other RNAs and prevent them from being used to make proteins. This gives the cell yet another layer of control over how the information in genes is used—like additional sets of brakes in a car.

Researchers suspect that the small molecules play important roles in a wide range of processes such as guiding the development of cells into specialized types or triggering cell death. But the tiny size of microRNAs has made them hard to identify in the genome and hard to analyze in cells. Even so, scientists have found clever ways to get a look at their functions. In 2008, Nikolaus Rajewsky, a professor at New York University and at the Max Delbrück Center in Berlin, Germany, carried out two groundbreaking studies to measure their impact on cells.

MicroRNAs bind to other RNAs and block their translation into proteins in two ways. In the first, they signal that the RNA should be destroyed by a molecular machine called the Dicer complex. One of Rajewsky's projects removed the main component of this machine—a protein also called Dicer—in immune system cells called B cells. Then his lab compared what proteins were made in these cells to control cells that had a normal Dicer complex. "We found 411 different molecules that were being produced at significantly higher levels in the experimental animals," Rajewsky told the author. "The production of dozens of these molecules would normally have been prevented because Dicer would have chewed up their RNAs. Since we stopped that from happening, they slipped through."

MicroRNAs have a second way of blocking protein production without causing their targets to be destroyed. They bind to a messenger RNA and prevent ribosomes from reading its information. The goal of another study by Rajewsky's lab, published in the journal *Nature* in 2008, was to measure how much of each type of control happened in cells. They found that a single microRNA can directly "tune down" the amount of proteins produced from hundreds of different mRNA molecules. Often it uses both strategies to do so, blocking translation and signaling for the mRNA target to be destroyed. For individual molecules, the effect is more like a volume control than an on-off switch. A microRNA usually reduces the amount of a particular protein that gets made rather than completely blocking it. Usually, at least 25 percent of a given mRNA escapes control. But the effects are strong enough to have a significant impact on the cell. If "generic" cells are prompted to start making a microRNA that is normally found in a particular cell type, they usually start to take on the characteristics of that type.

RNA KNOCKOUTS

Scientists quickly saw that the principle behind microRNAs could be turned into a new method of controlling genes, possibly even as therapeutic tools. The knockout methods described so far work by switching genes on and off. Some of the newest methods block genetic information contained in RNAs rather than genes. This can give scientists an even finer level of control over gene activity. For example, knocking out a gene in a specific tissue means that no RNA will be produced at all. It would not be possible, therefore, to study the differences in behavior of RNAs from the same gene that have been spliced in different ways. RNA knockouts make this possible.

The strategy is to create artificial RNAs called *small interfering RNAs* (siRNAs), which are artificial and often more powerful versions of microRNAs. They work the same way, by triggering the destruction of mRNAs or preventing their translation into proteins. Multiple genes can be shut down as well, because several siRNAs can be introduced into cells at the same time. A consortium of laboratories in Europe is currently using the method to look for molecules involved in cancer. The strategy of the project, called "Mitocheck," is to shut down successive RNAs and watch what happens to the cell. If the loss of a molecule causes the cell to reproduce at the wrong time, or at a much more rapid pace than usual, it may mean that the molecule is involved in cancer. The next step is to collect tissue samples from patients and check whether there have been mutations in these genes.

This particular project involves single cells, but siRNAs are also being used in place of other knockout methods in animals. Instead of removing a gene, scientists insert an artificial gene to produce an siRNA that will bind to an existing RNA. It can be made to appear in specific tissues using conditional knockout techniques. That requires engineering sperm, egg, or embryonic cells.

Since the manipulation of human embryos is considered unethical, researchers are particularly interested in finding ways to deliver siRNAs to tissues in fully grown organisms. There have already been some promising early attempts to use the method in therapies on humans suffering from an eye disease called macular degeneration.

A major problem with all therapies involving siRNAs is that the molecules break down rather quickly in cells, so their effects are usually short lived. Before siRNAs become a part of the doctor's toolbox, a great deal of further work will be necessary in animals. In 2003, geneticist Beverly Davidson and her laboratory at the University of Iowa tried this technique with mice suffering from a form of Huntington's disease. This severe mental disorder is caused by mutations in the huntingtin gene, leading to a malformed protein that forms huge clumps that cannot be broken down. Eventually, this kills cells in a crucial part of the brain, leading to a breakdown of the nervous system. People lose control of their muscles, develop dementia, and eventually die. Huntingtin is necessary to the function of the brain, so therapies cannot simply aim to shut it down. Using mice with the same genetic defect, Davidson's lab showed that slowing down the production of the protein with siRNA dramatically slowed down the development of the disease. Researchers now hope to try a similar approach in human versions of the disease.

The first successful clinical trial of siRNAs was carried out in 2005 by Sirna Therapeutics, a pharmaceutical company based in Boulder, Colorado, on patients suffering from macular degeneration. People with this disease lose their eyesight because of the death of vision cells called rods and cones or because blood and proteins leak into the inner lining of the eye. Doctors treated the patients with an siRNA to block the production of a protein that plays a key role in the disease. All of the patients showed improvements over the course of 157 days with no signs of side effects. Currently, more clinical trials are planned in hopes of using siRNAs to treat other diseases.

MOLECULAR MACHINES

The behavior of proteins has turned out to be another example of incredible complexity within the cell. When first discovered, single molecules seemed incredibly powerful and important, like the stars of films. But just as a director, cameraman, and dozens of other specialists are needed to get an actor onto the screen, proteins also require a huge amount of technical support to do their jobs. They usually work in "machines" of various sizes, the largest probably containing more than a hundred molecules. These are continually being disassembled and rebuilt to do new things.

In 2005, Anne-Claude Gavin, Giulio Superti-Furga, and their colleagues at the company Cellzome in Heidelberg, Germany, worked with scientists from the nearby European Molecular Biology Laboratory to capture the first complete snapshot of all the machines at work in a eukaryotic cell. They discovered 491 machines in yeast; human cells probably build at least six or seven times that number.

Most of the protein machines in yeast are found in a similar form, using related proteins in human cells. This is strong evidence of evolution and gives important insights into how it works. The most important machines arose in an ancient cell. The components were passed down to humans, other animals, and plants, where they assembled into similar machines. Along the way, there has been a lot of fine-tuning: Machines have acquired new functions through the addition of new parts or slight changes in their shapes. Many have a "snap-on" structure: The cell prefabricates and assembles most of the parts ahead of time, often leaving a few pieces to be made when the machine is needed.

Building machines requires precise timing in the production of thousands of molecules, and the completion of genomes has brought along new methods to watch how this happens. One of the most important techniques is the microarray, or the DNA chip, developed in 1994 by Patrick Brown of Stanford University and the California-based company Affymetrix. The technology acts as a surveillance system that can detect whether cells have produced RNAs from particular DNA sequences.

DNA chips compare the gene activity of cells—for example, a healthy cell and one that has become cancerous or cells that have specialized into different types—to try to discover differences in the behavior of genes. A scientist extracts RNAs from both kinds of cells and tags them with different fluorescent markers. Then he or she exposes them to the DNA chip, which traps them. The effect is like going to a football stadium and trying to decide where the fans of each team are sitting. That will be easy if a lot of people have come to the game in team colors and are sitting together. A DNA chip compares the "cancer team" (suppose they are dressed in red) to the "healthy team" (in green). Some parts of the stadium will look mostly red—in cellular terms, this means that the cancer cell produces more of a molecule than the healthy one. A part that looks mostly green means that the healthy cell produces more of a particular molecule. Some parts of the stadium may have an equal mix of colors (the molecule is active in both cells), and others may be entirely empty (meaning the gene is not used by either type of cell).

Experiments with DNA chips reveal when and where the cells in an organism's body activate particular genes (including the components of specific machines) and watch how that behavior changes during disease. Switching on a single gene may trigger an avalanche of responses from other genes, with effects such as telling it to divide, altering the cell's form and behavior, or prompting its development into a new type. Disrupting any of these processes can lead to diseases such as cancer, in which cells forget their identities and functions, reproduce at the wrong time, and go on strange migrations through the body.

The discovery of so many protein machines has changed how scientists look at genetic diseases and other types of illnesses, such as cancer. Therapies may need to focus on fixing machines rather than trying to replace single, defective molecules—the way a clever engineer may be able to repair a motor by improvising something new if the original part is no longer available.

DNA chips and other technologies that can monitor the entire genome have given scientists their first look at the true complexity of biological processes. But they are only a beginning. The next step is to understand how organisms coordinate the activity of hundreds of millions of cells in organs such as the heart and the brain, and then to understand how those organs work together. Only then will the influence of genes on people's behavior and lives be truly understood. Today's scientists are developing a new genetic toolbox that has already shed a bit of light on the issues. The following sections discuss the state of the art and outline some of the questions that tomorrow's biologists will face.

CONDITIONAL MUTAGENESIS

The first techniques to knock out genes completely removed them or eliminated their functions in all of an organism's cells. Sometimes this had no effect at all, because other molecules were able to compensate for the loss. In other cases, it caused such severe defects that an embryo never developed, which made it impossible for scientists to understand the function of a particular gene. Another problem with all-or-nothing mutagenesis stems from the fact that the same gene may be needed at different times to do different things in various types of cells. For example, a protein called PS1 seems to act as a switch for different types of functions: It is needed to pass important signals that tell some types of cells to grow and develop. At other times and places in the body, it is involved in *apoptosis,* a type of cell suicide that is necessary as tissues form. If the gene for PS1 is removed, these processes can no longer be controlled.

It is not surprising that proteins have multiple functions or even tasks that may seem contradictory. The set of human genes evolved from a much smaller set in ancient ancestors that had much simpler bodies; the earliest consisted of a single cell. Sometimes new genes arose, but often evolution worked by copying existing ones and tinkering with their functions. Just as electric devices have some of the same components, multiple systems in the body rely on common proteins that have adapted to different tasks. So yeast, which is a single cell, contains proteins that now help build brains, eyes, and other highly complex organs in animals. Within an organism, nerve cells and muscle cells (just to give one example) use the same molecules to accomplish different things.

These multiple functions cause problems in a complete knockout. If a gene is crucial during the first stages of life, removing it may kill the embryo, making it impossible to study later functions. Likewise, inserting a foreign gene might affect several types of cells in different ways. So scientists began looking for ways to gain more precise control over gene activity.

In the mid-1990s, Klaus Rajewsky, Frieder Schwenk, and their colleagues at the University of Cologne in Germany found a way around this problem with the invention of conditional mutagenesis. Their method relies on the fact that genes have a complex structure (described in chapter 3) including repressors, operators, and various other control elements. Rajewsky's lab built genes with artificial switches that gave the scientists control over when and where a gene is shut down in an organism.

As with the use of recombinases (described in chapter 4), the first of these methods was based on borrowing molecules from bacteria. An initial discovery was an enzyme in bacteria called Cre that recognizes patterns in DNA called loxP sequences. If Cre finds two of these sequences in DNA, it binds to the sites and draws them together, making a loop of the DNA that lies between them. This looped sequence is cut out, destroying a gene or any other information that it contains (such as regions that control a nearby gene). The cell then repairs the break by gluing the cut ends together. Thus, the first step in creating a conditional mutant is to build an artificial gene centered between loxP sequences.

The key to the method is that DNA is destroyed only in cells that produce both Cre and loxP sequences. If they are active in all of an organism's cells, the effect is like a complete knockout. Since the whole idea behind conditional mutagenesis is to avoid this, Rajewsky and his colleagues had to find a way to activate Cre only in particular types of cells. The solution was to find other genes that were used only in specific tissues or cell types. Some proteins are produced only in the brain, for example, because they are controlled by DNA sequences that are used only in certain types of neurons. By combining Cre with such a control region, scientists could ensure that it, too, would become active only in the brain. The same technique could be used to study genes in any other tissue, providing a unique control region could be found.

Further refinements now allow scientists to determine the time as well as the location in which Cre becomes active. This is accomplished by attaching yet another switch to the Cre gene, such as a receptor called LBD. This molecule becomes active only in the presence of a hormone. Since animals do not naturally produce the hormone or obtain it through their diets, Cre remains inactive until the desired time, when the animal is given the hormone in food or an injection.

In the 1980s, researchers learned to apply the tools of genetic engineering to animals such as mice. This image shows a laboratory mouse in which a gene affecting hair growth has been knocked out (left) next to a normal lab mouse. *(National Human Genome Research Institute)*

Introducing Cre and the loxP-marked gene into mice is accomplished by genetic engineering in separate steps. They are inserted into different strains of mice, which are then mated to produce offspring having both Cre and the targets. While this means waiting at least two generations for a mouse that has both elements, it also permits scientists to mix and match various Cre strains with mice that have different genes marked by loxP. If the same protein is needed in the brain and the kidney, for example, and its gene has been tagged with loxP, scientists can mate the mouse with one Cre animal to test its functions in the brain, and another to see what it does in the kidney.

Ideally, researchers would like to have a strain of mouse in which each gene is surrounded by loxP elements, and other strains that express Cre in each tissue and cell type. Theoretically, this would allow them to test the function of every gene in every kind of tissue. It would be an enormous amount of work, as mice have at least 13,000 genes and at least several hundred different cell types. Yet the usefulness of the mouse in creating human disease models has convinced many researchers that doing so could be worth the effort. This has encouraged scientists to start creating Cre zoos—collections of animals expressing Cre in different tissues. These animals are commonly shared by different labs, saving time and reducing the number of animals used in research. Centralized collections of mouse strains have been established at Jackson Laboratories in Maine, the European Mutant Mouse Archive near Rome, Italy, and elsewhere.

In the meantime, other molecules have been developed to work like the Cre-loxP pair. One example is a yeast enzyme called FLP, which recognizes target sequences called FRT. FLP was originally difficult to use in warm-blooded animals, because yeast normally grows at lower-than-body temperatures (84°F, or 30°C). More recently, scientists have altered the molecule to perform better at higher temperatures. This means that strains of mice containing both Cre and FLP systems can also be mated to give researchers control of several genes in a single animal. They can be combined, for example, to see what happens when different genes are shut down in sequence.

Studying such animals will not solve all questions about the functions of genes in humans or even in mice, because everything that happens in cells and organisms requires the collaborative efforts of many genes. Scientists dream of someday

Other Methods of Controlling the Output of Genes

The first knockout methods were crude, because they removed genes or their control elements from an entire organism. Conditional mutagenesis was much more precise, because it could be used to remove genes at a specific time and place. But even this level of control is not precise enough to completely reveal the functions of a gene. If the RNA transcribed from a gene can be spliced in different ways to make different proteins, which is usually the case, a conditional knockout removes them all in a particular type of cell. siRNAs can be designed to influence specific RNA molecules and leave others intact. But these molecules are so small that they often hit unintended targets, binding to more than one RNA and affecting too many genes. Recently, scientists have been developing other methods of engineering RNAs that affect single molecules.

The enzymes originally used in genetic engineering have come from bacteria; in the meantime, scientists have branched out to look for tools in some of the most exotic organisms on the planet. The laboratory of Glauco Tocchini-Valentini near Rome, Italy, has been building a new type of RNA control mechanism from single-celled organisms called archaea, which live in volcanic vents on the ocean floor, in geysers, and in other extreme environments. The first such organisms were discovered about 50 years ago and were originally thought to be bacteria. Closer study showed that they were another form of life, distant from both bacteria and plants and animals. In fact, the earliest cells on Earth were more like archaea than anything else alive today.

(continues)

(continued)

Life in extreme environments has led to the evolution of molecules with unique properties in archaea. One of these, called a tRNA endonuclease, works a bit like DNA recombination (described in chapter 4) except that the tRNA endonuclease targets RNA molecules. When it recognizes a particular shape, the molecule cuts it out and rejoins the loose ends. Glauco Tocchini-Valentini and his colleagues at the Italian National Research Center are using this feature to turn it into a powerful tool for genetic engineering.

The particular shape that the enzyme recognizes is called a BHB—short for a folded shape called a bump helix bump—and by making molecules take on this shape, the researchers can use the endonuclease to shut them down or change their functions in other ways. RNAs sometimes acquire this shape if they contain different regions with nearly complementary sequences that bind to each other. The helix consists of a short stretch of four nucleotides that bind; on either end are bumps that form because a few letters of the sequence do not bind. The effect is a little like gluing two strips of paper together and sliding them before they have dried: If the glue has not been spread evenly, a wrinkle may form.

When the tRNA endonuclease feels this shape, it makes a cut on the outside of each bulge, releasing the fragment in the middle. That part gets destroyed, and another enzyme comes around to glue the broken ends of the RNA back together. Any part of an RNA can make the BHB shape if scientists add a partner molecule to the cell with just the right sequence. Then they add the cutting enzyme, which removes a vital part of the RNA and shuts it down. This is more specific than a gene knockout, because only cells that contain both the RNA and the

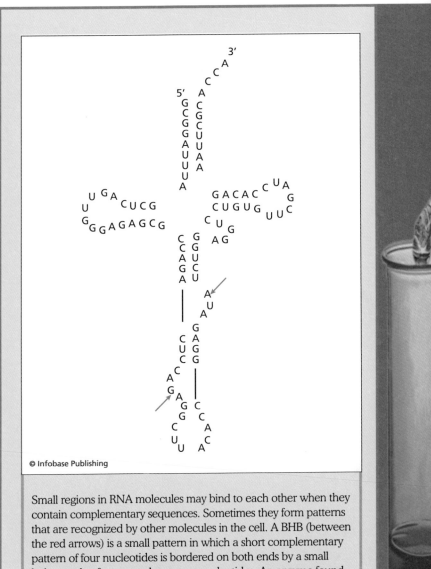

© Infobase Publishing

Small regions in RNA molecules may bind to each other when they contain complementary sequences. Sometimes they form patterns that are recognized by other molecules in the cell. A BHB (between the red arrows) is a small pattern in which a short complementary pattern of four nucleotides is bordered on both ends by a small bulge made of noncomplementary nucleotides. An enzyme found in one-celled organisms called Archaea recognizes this pattern, cuts it out, and rejoins the ends. Genetic engineers are learning to use this system to change the behavior of RNA molecules.

(continues)

(continued)

enzyme are affected. And the process does not destroy every molecule produced by the gene, only the exact RNA that contains the BHB shape. Thus, if they wish, scientists can target a single spliced version of an RNA, rather than every protein form encoded in a gene.

Another use of the molecular tools is to fuse separate RNAs into a new molecule (the technical term for such hybrids is chimera), which happens when two different RNAs overlap in a BHB shape. The enzyme cuts out the BHB, and when the broken ends are rejoined, the separate RNAs have been grafted together. They are then used to make a chimeric protein. Such hybrid molecules are interesting, because they can be used to make or break theories about proteins' functions. If researchers believe that a protein works because it occupies a certain position in the cell, they can attach it to another molecule which is attached somewhere else. The protein may have everything it needs to do its normal job, but it cannot get to the right position in the cell.

Very precise techniques like this are particularly needed when it comes to investigating the effects of several genes at the same time. The difference is like using a pair of tweezers, rather than a magnet, to pick out one key from a big pile. The magnet will likely work, but it will pull in too many keys. Knockout techniques that affect too many proteins will not give scientists the fine level of control they need to thoroughly understand systems in cells and how they go wrong in disease.

developing research animals with switches on every gene, giving them control of many genes at the same time and allowing them to investigate complex patterns.

MOLECULAR MEDICINE AND GENE THERAPIES

Until very recently, the greatest worldwide threats to human health were infectious diseases. Modern sanitation, antibiotics, and vaccines have changed that situation in many places but not everywhere. Malaria, AIDS, and other infectious diseases continue to decimate huge regions of the globe, and new plagues will continue to arise and cut swaths of destruction before cures are found. But in developed and developing countries, the major killers have become old age conditions such as cardiovascular and metabolic diseases, cancer, and neurodegenerative diseases. The culprits in these cases are partly our own genes, partly the environment.

The last revolution in medicine could take advantage of the body's own defenses in combating disease. Vaccines stimulate the existing immune system, which arose under the pressures of evolution: Natural selection works to protect organisms as long as they reproduce and care for their young, and viruses and bacteria attack the young. Animals thus developed a sophisticated immune system to protect them from such diseases. Modern medicine was born when Louis Pasteur, Robert Koch, and researchers of the late 19th century discovered that infectious diseases were caused by bacteria, parasites, and viruses, many of which could be treated with antibiotics and vaccines. But humans do not have natural defenses against today's major health threats, which arise late in life, often as a result of normal aging processes. Curing these diseases will require a new type of medicine that addresses their causes.

When a disease arises because a person has inherited a defective copy of a gene, scientists hope to find a strategy to replace it with a healthy version of the molecule. There are several hurdles that have to be overcome: The gene has to be delivered intact to the cells where it is needed, it has to work there, and it has to avoid being rejected by the immune system. The latter problem is particularly difficult because cells have evolved sophisticated defenses to prevent the entry of foreign genes.

One strategy that is being tried is to remove a patient's own cells, equip them with the healthy gene, and reimplant them in the hope that the body will still accept the cells, even though they have been altered, as its own. In the best scenario, the cells will divide and supply the body with a long-term supply of the healthy molecule. The first human experiment of this type was conducted in 1990 on a four-year-old girl named Ashanti DeSilva. She suffered from a rare genetic disease caused by a defective version of a gene called adenosine deaminase (ADA). The disease made her white blood cells unable to protect her from infections. Most people with the condition die at an early age. W. French Anderson of the National Heart, Lung, and Blood Institute and two researchers from the National Cancer Institute, Michael Blaese and Kenneth Culver, removed white blood cells from Ashanti, supplied them with a healthy version of ADA, and reimplanted them. The cells now worked properly, allowing Ashanti to lead a normal life, but they failed to divide and make new copies of themselves. She continues to receive the therapy, which is supplemented by doses of a drug containing ADA protein.

It may also be possible to deliver healthy genes directly to cells without having to remove and alter them. Viruses often manage to deliver foreign DNA, RNA, and proteins into cells, which suggests that they might be used as delivery vehicles for healthy genes. The strategy would require rebuilding an existing virus by keeping what it needs to invade cells, removing anything that might be harmful, and packing it with new DNA or RNA. Once this engineering were done, the virus would be allowed to infect a patient who needed the healthy molecules.

Before scientists carried out such experiments, they had to have an exhaustive understanding of how viruses were built and functioned. Viruses mutate very quickly and might become dangerous, even infectious. New strains of the flu arise almost every year when an animal virus mutates and can suddenly "jump the species gap" to infect humans. Something similar could happen with a virus that had been genetically engineered. But more realistic than the horror scenario of a global epidemic is the potential danger to individual patients, which cannot be completely foreseen.

The first experiments were carried out using viruses that easily infect humans but almost never cause serious diseases. About a quarter of *gene therapy* experiments have involved a strain of the adenovirus whose worst effect is usually a mild cold. Researchers weaken the virus by deleting some of its genes, then replace them with the healthy human genes needed by the patient. Even after extensive animal experiments, the risks cannot be completely eliminated. Most viruses are specific to one or two species, and the factors that prevent them from jumping from mice to humans cannot be adequately tested in animals.

A few early successes were followed by frustrating failures: One patient in a study of gene therapy for cystic fibrosis suffered inflammations in the lungs. Then, in 1999, an 18-year-old named Jesse Gelsinger died as a direct result of gene therapy. The case drew a great deal of attention from the media, and the doctors responsible for running the experiment were sharply criticized for not having noticed problems in earlier versions of the experiments.

Jesse suffered from a rare disease called ornithine transcarbamylase deficiency (OTC), the result of a defective protein. This molecule normally helps the liver break down ammonia, which is naturally produced by the body but which is also toxic. If the liver cannot rid the body of it, the ammonia travels through the blood to the brain, leading to a coma and death. Most victims die as infants, but a new therapy combining diet and drugs has prolonged some patients' lives. Jesse had a mild version; he stayed alive by following a very strict diet and taking dozens of pills every day. When the chance arose to participate in a new gene therapy trial, he and 18 other people suffering from the disease jumped at the chance. They knew that the experiment would not offer a permanent cure, but it might ease some of the symptoms, and what was learned might help save lives in the future.

There were already signs that things might go wrong during trials with animals. More than 20 experiments with mice had gone well, but three rhesus monkeys died when they received a very strong dose of the virus, much higher than the doctors planned to use in the study with humans. An autopsy revealed that their blood had clotted abnormally, and their livers were severely inflamed. Without knowing exactly why this had occurred, there was a risk that the same thing might happen in humans, and two doctors reviewing the case recommended canceling the study. Others, however, argued that the procedure was safe enough to try in adult humans if dosages were carefully controlled. And working with adults had to be tried before moving to babies who suffered from severe forms of the disease.

Jesse received a stronger dose of the virus than most of the other patients and within a few hours had developed a fever and upset stomach. Doctors were not overly alarmed, because other patients had also experienced slight infections. The next morning, however, things took a turn for the worse. Jesse's liver seemed to be failing. Over the course of the next three days, his organs shut down one by one. Finally, four days after receiving the treatment, he died. The Food and Drug Administration and the NIH immediately shut down all adenovirus therapies pending an investigation of the case.

DNA VACCINES AND T-CELL THERAPIES

Some genetic diseases—such as those of Ashanti DeSilva and Jesse Gelsinger and DMD—are inborn and affect every cell in a person's body. Others, such as cancer and many autoimmune diseases, arise from mutations that occur in particular cells during a person's lifetime. They happen because of mistakes that occur naturally as DNA is copied, exposure to carcinogens such as cigarette smoke, and other reasons. The genes that are responsible are different from patient to patient, and often several genes are involved, so it will probably not be possible to cure them by trying to repair single molecules. Instead, researchers hope to train immune system cells to attack and destroy disease tissue the way they deal with viruses or bacteria. Two of the strategies that are being tried are DNA vaccines and T-cell therapies.

Most vaccines work because they contain foreign molecules from a virus, which sounds an alarm in the immune system

Personalized Medicine in the Genome Age

Although Jesse Gelsinger's case and a few others reveal the need for caution in the early days of gene therapies, several other remarkable studies show how promising this new type of medicine is likely to become. A good example is Duchenne's muscular dystrophy (DMD), a disease that affects one of every 3,500 newborn boys. DMD is caused by various mutations in a gene called dystrophin, which is important in building cardiac and skeletal muscle. The first symptoms usually appear when the boys are small children. Their muscles begin to weaken, and this condition becomes progressively worse. Eventually, the muscles that serve the lungs or heart usually fail, leading to death.

In February 2008, Leland Lim and Thomas Rando, two physicians for the Veterans Affairs Palo Alto Health Care System, reviewed current approaches to treating DMD in an article in the journal *Nature Clinical Practice*. Studies have shown that about two thirds of people with the disease have major problems with the gene—either a large part of the DNA sequence is missing, or a large block is repeated. In other cases, the problems are smaller, affecting just a few of the base pairs in the gene. The latter defects are similar to those found in a related disease called Becker's muscular dystrophy (BMD), which usually has milder symptoms, but which can also be fatal for the same reasons over a longer period of time.

The most severe mutations cause the body to build dystrophin proteins that are missing large modules that are needed so it can bind to other molecules and behave properly. The smaller mutations usually change the code

(continues)

(continued)

of dystrophin mRNA so that it is destroyed by the nonsense-mediated mRNA mechanism (NMD), described earlier in this chapter, before the mRNA can be used to make proteins. DMD and BMD are different, because the mutations have different effects. In BMD, dystrophin protein is still made, but the mutations make small changes in its shape or chemistry so that it does not work as well, or they may prevent the body from making enough of the molecule.

Lim and Rando describe two main approaches being used to treat the diseases. The first is gene replacement therapy, in which scientists try to deliver a healthy copy of the dystrophin gene to muscle cells. This is done using viruses (which have been rebuilt to carry the gene), or by injecting DNA directly into the muscle, hoping that cells will absorb it and use it to make healthy RNAs and proteins. A third method is to remove patients' cells, engineer them to carry a working form of dystrophin, and reimplant them in the muscle. All of these strategies have had promising results in experiments with mice, but several major challenges have to be overcome before they can be tried in humans—for example, getting the genes into enough cells and avoiding immune reactions.

The second main approach is called gene modification, and the idea is to use proteins or small bits of DNA to make changes in a person's existing dystrophin gene. How this will be done depends on the type of mutation a person has. The goal, say Lim and Rando, is to make cells handle DMD-type mutations the way they cope with BMD mutations—in other words, to turn the most serious form of the disease into a weaker form.

The authors cite several strategies that are being developed to accomplish this. One involves an antibiotic called gentamicin. This substance causes mistakes in the way ribosomes read mRNAs to build proteins, as if a teacher decided to grade papers with the help of a spell-checker and accidentally used a British version of the spell-checker rather than the American one. This might lead the teacher to give a bad grade to a good paper. In bacteria, gentamicin is usually harmful—it kills bacteria by disrupting the production of proteins. But in DMD, the problem is that the mRNA "text" really does have spelling mistakes. Scientists hope to use gentamicin like a super hacker that breaks into the spell-checker and tells it to overlook those mistakes. This will not make the cell produce healthy dystrophin, but the version that it makes will not be as bad.

Gentamicin is not harmful to healthy human cells and had already been approved to treat infections in humans, so Luisa Politano and her colleagues at the Second University of Naples, Italy, carried out a clinical trial in four patients with DMD to see if gentamicin would allow their bodies to produce dystrophin. Politano and her colleagues found various amounts of the protein in muscles of three of the four patients. The treatment is not a cure; it needs to be improved before patients' bodies build enough dystrophin to affect the course of the disease. But it shows that the strategy might be useful for some people. It does not work for all patients because the "hack" works only for particular spelling mistakes. But it may be possible to use other antibiotics that work in a similar way to treat other mutations. This will permit personalized gene therapies, custom made to fit the individual mistakes in a patient's dystrophin gene.

when it comes to the attention of white blood cells called B cells and T cells. Both types originate in a person's bone marrow. B cells produce antibodies—Y-shaped proteins that decorate their surfaces. These molecules are created in an unusual way, by cutting and pasting genes together in random arrangments and using them to create antibody proteins. The result is that the body can probably produce about 10 billion different types of antibodies—as if a locksmith were to take millions of blank keys and cut them in random shapes. Chances are, one of them would fit and open the door of a particular house. In this case, an antibody needs to fit an antigen—a protein on the surface of a bacterium, virus, or other invader. If the antibody succeeds, the cell that carries it can trigger a reaction from the immune system.

This usually happens as a cooperative effort involving both B and T cells. T cells do not have antibodies, but they have receptor proteins on their surfaces that recognize foreign proteins in a similar way. A full-blown immune reaction usually requires that both a B and T cell recognize an invader and then meet each other. When that happens, the T cell activates the B cell, which begins to divide very quickly. This creates a huge number of identical daughter B cells that make billions of copies of the same antibody and secrete it. The free antibodies glue themselves onto the surface of the pathogen, acting as a sort of alarm beacon. This summons huge cells called macrophages that swallow the foreign object and break it down. (Anyone who has seen the movie *Fantastic Voyage* will remember macrophages as the huge cells that swallow a miniature submarine.) The system does not always work because some infectious agents reproduce very quickly or manage to disguise themselves.

If the immune system has defeated a pathogen once, it produces a special type of B cell called a memory cell. If the virus or parasite returns, this cell can be turned into a factory to produce huge amounts of the antibody again very quickly. Vaccines mimic this effect by preparing the system ahead of time, teaching the immune system to recognize a protein from the virus or second virus that is so similar, the body cannot tell the difference. Vaccines have had a huge impact on human health and society, helping humanity to wipe out some of its worst diseases, and have even changed the course of history. One of the earliest uses of something like vaccination may have helped the United States win the Revolutionary War. George Washington decided to protect the Continental army by inoculating them with pus taken from the sores of smallpox victims. While this procedure was dangerous, it greatly lowered the risk that a person would catch and die from smallpox. Without the treatment, the disease might have run rampant through Washington's troops and made the army too weak to fight the British. A few years later, Edward Jenner (1749–1823) developed a safer method using the cowpox virus to vaccinate people against smallpox.

Many viruses do not have a close relative that can be used as a vaccine, so researchers have to produce one from the dangerous form, either by weakening or killing it. Yet there are still risks—viruses altered in this way sometimes cause infections or side effects. And it still has not been possible to create vaccines against bacteria, many parasites, and even some viruses that evade the immune system by changing the molecules on their surfaces.

On the other hand, some parts of the genomes of bacteria and parasites rarely change. This led scientists to wonder whether it might be possible to create a DNA vaccine—to teach the body to recognize specific foreign DNA sequences and destroy them, in the same way that foreign proteins call up the cell defenses. The method involves finding a DNA sequence that is present only in the invader and not in human cells. This was not possible until the human genome was finished, because there was no way to scan the entire sequence for possible matches. Now that this can be done, scientists can take a unique bit of DNA, copy it, and inject it into muscle. Sometimes cells take it up and use it to produce proteins. The immune system recognizes these as foreign and attacks the invader. It might be possible to produce a stronger effect by using several unique genes from a pathogen; this could help immune cells recognize viruses or bacteria whose surfaces have changed. Other advantages are that doctors would no longer have to inject a complete (weakened or dead) virus into patients, which would eliminate many of the side effects of vaccines.

The idea is appealing, but will it work? Early trials failed to provoke the immune system into responding strongly enough to defeat the disease. But a few recent studies have had more success. In 2004, scientists at the U.S. Army Medical Research Institute of Infectious Diseases in Maryland injected rhesus macaque monkeys with four genes from the smallpox virus and made them immune to monkeypox. And in 2006, patients responded positively to an experimental DNA virus against the bird flu.

T-cell therapies work more like the gene therapies described in the last section and are currently being used in cancer therapy trials. The idea is to fool T cells into thinking that a tumor cell (which has been produced by the body) is foreign. First, scientists have to identify a tumor marker—a molecule that appears only in cancer cells but not in healthy ones. All cells that bear the marker will become targets and be destroyed, so it has to be unique to tumor cells. In 2008, Ute Stein, Peter Schlag, and their colleagues at the Charité University Hospital in Berlin, Germany, found such a marker in colon cancers. Cancer cells that are about to become metastatic—wandering to other parts of the body to build new tumors—produce a protein called MACC1.

The next step is to create a receptor protein that can recognize MACC1 or whatever marker has been identified. This is inserted into T cells taken from the patient's body. The cells reproduce in the laboratory and are then transferred back into the patient. If everything goes well, the T cells will train the immune system to recognize and attack the tumor.

This strategy also has risks. The surfaces of cells contain thousands of different types of molecules, and the T cell might have extra receptors that target other proteins in the body and cause autoimmune reactions. One solution is being developed by Wolfgang Uckert's group at the Max Delbrück Center for Molecular Medicine in Berlin. When the scientists create a new T-cell receptor, they add a tiny bit of extra code that acts like a self-destruct button. If things start to go wrong and transplanted cells lead to an autoimmune problem, the patient can be given antibodies that recognize the extra code. This causes the T cells to be destroyed by other parts of the immune system.

The success of the first gene therapies in the early 1990s raised hopes among the public that molecular biology and genetic engineering would quickly lead to new drugs and cures. Most scientists have had a more cautious view. While these techniques have already led to powerful new diagnostic tools to detect and analyze diseases, it will still be many more years before gene therapies, DNA vaccines, and T cell therapies will be standard tools in the fight against cancer and other diseases. But most scientists are confident that this will happen. Modern medicine was ushered in when Koch, Pasteur, and other scientists recognized the causes of infectious disease. The new understanding of life that is emerging in the age of genomes has already had a dramatic effect on medicine, and this trend will surely continue.

6

Ethics and Genetic Engineering

Over the last 150 years, science has created a new view of humanity (as a species produced by evolution), has dramatically changed people's lifestyles (through discoveries that have revolutionized travel, communications, and many other technologies), and has continually raised unexpected challenges and issues. Scientific discoveries often open uncharted ethical territories for which no clear answers can be found in the religious or cultural traditions that used to be the main source of guidance in resolving moral problems. Lawmakers need ethical advice as they try to structure society to provide peace and freedom for citizens, but where should they turn to get it? In countries such as the United States, whose citizens practice every major religion in the world and where religious freedom is a fundamental right, it is difficult to find common ground. This is where ethics—the study of moral ideas with the aim of helping people decide how to behave—comes in.

Ethical questions involve decisions about what is good, what is desirable, and what will be in the best interests of society. These questions affect everyone and usually lie outside the expertise of scientists. The role of researchers—and of books like this one—should be to ensure that decision makers and the public have clear, accurate information about how research is done; to elucidate what potential risks and benefits it may bring; and to point out ethical issues that may not be obvious to nonexperts. Only then will people be able to make good decisions about how to regulate today's science in the interests of future generations.

The field of genetics has raised significant ethical issues for several reasons. First, it reveals new information about people's genes and their family histories, raising issues of privacy. Other issues arise as genetic methods are used to alter the genomes of plants and animals, potentially even of humans. If genetic science achieves what most researchers expect, it is likely to find cures for cancer and genetic diseases and extend the human lifespan, which will have an important impact on society. This chapter presents some of the most important questions that have arisen and some of the ways that scientists and others are coping with them.

REPRODUCTIVE CLONING

In early 2004, geneticist Hwang Woo-Suk announced that his laboratory at Seoul National University in South Korea had "successfully culled stem cells from a cloned human embryo through mature growing process in a test tube," according to a CNN press release from February 13, 2004. The announcement of the first successful cloning of a human triggered panicked headlines and drew researchers from all over the world to Asia, where they hoped to learn the techniques. Hwang Woo-Suk's claims soon turned out to be a case of scientific fraud, but another group has succeeded in producing human embryo clones. In early 2008, Samuel Wood, head of a California company called Stemagen, announced that nuclei extracted from his own skin cells had been successfully used to create embryos that survived five days. Wood's accomplishment was immediately both praised and strongly condemned in the popular media. For example, Monsignor Elio Sgreccia, a spokesman for the Vatican, called the work the "worst type of exploitation of the human being," and said it "ranks among the most morally illicit acts, ethically speaking." His comments were reported in the January 18, 2008, issue of the online news service *Reuters Health*.

Scientists use the term *cloning* to refer to the production of genetically identical segments of DNA, cells, or organisms from

Dolly the Sheep

In 1996, cloning of the first mammal—a sheep named Dolly—brought sensationalist headlines and concerns that humans would be next.

a single parent. In science fiction and the media, it usually refers to "copying" a person, usually an adult, all the way to the clone's birth and beyond. Scientists refer to this as *reproductive cloning*. At the time of this book's publication, there have been no confirmed cases of successful human cloning. But technically, it will probably be possible very soon. This has revived a number of ethical concerns about the use of science in human reproduction.

Intense debate about cloning began in 1996, when researcher Ian Wilmut (1944–), at the Roslin Institute in Scotland, announced that a sheep had given birth to its own clone, named Dolly. The technique used by Wilmut (and later Wood) is called somatic cell nuclear transfer. Scientists remove the nucleus of an egg and replace it with the nucleus of an adult cell. The egg is then stimulated by electricity or chemicals to grow into an embryo. In Dolly's case, researchers began with a cell from the mammary gland of a female sheep. This made her an identical twin of her mother—except that she was born many years later.

Dolly demonstrated that mammals could be cloned this way, but the price was high. Producing a single living sheep required 277 eggs. Only 30 of them were able to divide after the transfer of the foreign nucleus. Nine of these triggered a pregnancy (the others did not become implanted in the mother's uterus), and only one survived to birth. As she grew, Dolly suffered continual health problems. Some scientists have interpreted these as signs of accelerated aging, although this is disputed by Dolly's caretakers.

Trying to clone a human being would probably be even more difficult. But suppose the technical problems could be solved. Would human cloning become desirable? What impact would it have on society? Glenn McGee, a bioethicist at the Center for Bioethics of the University of Pennsylvania School of Medicine and editor in chief of the *American Journal of Bioethics,* says that society needs to resolve several questions before reproductive cloning can become a standard practice in human reproduction, including the following:

- Is cloning unnatural self-engineering?
- Will failures, such as deformed offspring, be acceptable?
- Will cloning lead to designer babies whose freedom is restricted in the future?
- Who is socially responsible for cloned humans?
- Do clones have rights and legal protection?

Lewis Wolpert, a developmental biologist at University College London, points out that all of these questions equally apply to in vitro fertilization—the creation of so-called test-tube babies— which has been commonly practiced for more than three decades. Since the birth of Louise Brown in 1978, the first baby born from an egg fertilized in a test tube, at least 3 million babies have been born using this procedure. Like cloning, this process requires the creation of many embryos that do not survive to birth, and there is an ongoing debate about whether scientists should be allowed to carry out research on cells that will die anyway. They could provide a source of embryonic stem cells that might be useful in treating diseases. Most of the debate centers around the questions, "At what point does an embryo become a human being, and what legal protection should it enjoy at different stages of development?" Those who believe that human life begins at conception—or earlier—obviously object to cloning and many other reproductive technologies that are commonly used today.

A fertilized embryo that is never implanted in a mother's womb has no chance of becoming a human being, and some argue that this makes its status no different than tissue samples taken from patients or other types of cells grown in a test tube. Yet others believe that even the potential to become life means that embryos should be protected. This issue is discussed in the next section, which deals with the making of clones as a source of cells for therapies.

Some of the concerns that have been raised about cloning seem to be responses to science fiction books and films that present a distorted view of cloning. In *The Sixth Day,* a science fiction film released in 2000, Arnold Schwarzenegger plays an adult who has been cloned. Somehow, the clone grows at an accelerated rate and acquires all the memories and behavior patterns of the original, to the point that no one can tell the difference between the two. Even the clone believes he is the original. In reality, people and their clones would be like identical twins of different ages who have the same genes but different memories and personalities and who would be influenced by the environment in different ways. Clones in films also frequently appear as the slaves or property of other people. This is strange considering that one twin does not belong to another, and a test-tube baby does not belong to a scientist who fused an egg and a sperm in a laboratory. These principles have led British biologist Lewis Wolpert to claim that cloning does not raise any new ethical issues. (He once promised a bottle of champagne to anyone who proved him wrong.)

Normal identical twins are an example of "natural" cloning, in which cells belonging to the same organism become separated very early in life. At this early stage of development, each cell is totipotent—it retains the ability to differentiate into every type needed in the adult body. But soon cells begin to specialize along different paths, because they activate particular subsets of genes. While each cell retains a copy of the entire genome, it uses the information in different ways, activating some genes and keeping others silent. Cells remember these states as they divide and differentiate. So the nucleus of a skin cell somehow knows that it is skin, and if it were implanted into an egg cell, it might no longer be capable of creating healthy cells of all the types needed to build a complete organism. The nucleus might also be old—as a person ages, DNA becomes damaged, and other changes might be carried along as it forms a new organism. The nucleus used to create Dolly was six years old at the time of her birth. Would she age normally, or would her development be more similar to an animal that was six years older? The study was inconclusive. Dolly suffered from arthritis, which is usually found in older animals, but the lung disease that killed her at the age of six may have had nothing to do with the fact that she was a clone.

Again, supposing that technical issues could be resolved, why would anyone be interested in spinning off identical twins of an existing person? The main reason to do this—rather than create genetically new individuals from the combination of a unique sperm and egg—is that clones could be useful. If scientists can stimulate the cells to develop into specific types of tissues or organs—which can be done in animals—stem cells could be harvested from a clone to replace a person's damaged tissues. Such stem cell therapies would avoid the problems of tissue rejection that normally accompany transplants. On the other hand, if the person suffered from a genetic disease, the cells would contain the same defects. It might be possible to correct them, which is one aim of therapeutic cloning, discussed in the next section.

Another proposal has been to save the DNA of very gifted people such as artists and scientists. Behind this type of thinking, too, is usually the idea that "reincarnating" the person at a later date might be somehow useful to society. But studies of twins have shown that two people with identical genomes can develop in very different ways, acquiring different talents and tastes. The same would be true of clones. Each newborn would be an individual who would develop a unique personality and skills. The clone would probably be born several decades later than his or her "parent," into a very different culture. So there is no guarantee at all that a clone will share the interests or talents of the original, even though society might have great expectations of him or her. Despite the horror scenarios presented in science fiction books and films, clones would not be able to read each others' minds or share memories.

THERAPEUTIC CLONING AND EXPERIMENTS WITH HUMAN CELLS

Hospitals and laboratories throughout the world use human cells in experiments. Usually, these are samples taken from tumors, organs, or other tissues with the permission of the patient. Another use foreseen for human cells is called therapeutic cloning. The idea is to harvest cells from an adult, not to make a copy of a person, but to create new cells that could be useful in fighting disease. For example, clones could be transformed into the precursors of cells that the body has a hard time replacing, such as muscle and neurons. If scientists learn enough about the genes that guide tissue development, they may even be able to grow complete organs in the laboratory.

Such cells would be useful in treating many diseases caused by the death of particular kinds of cells. Cell death is a natural process—red blood cells, for example, have a lifespan of about 120 days—and the body can replace many of them. It does so by drawing on adult stem cells stored in various places in the body. In humans, these are capable of replacing some types of cells and repairing some tissues, but not all. Mammals have lost the ability to rebuild major tissues and organs, but species such as salamanders can regrow entire limbs and organs that have been lost or damaged. The most powerful regenerators may be worms called planaria, which can be cut up into hundreds of pieces, each of which will develop into a full worm.

One way to fight degenerative brain conditions and other wasting diseases might be to implant fresh stem cells that find the damage and transform themselves into new neurons or other types of cells. This has been done successfully in animals, and it is what happens with bone marrow transplants. The marrow is a factory and storage facility for many types of stem cells. People with certain genetic diseases or cancers produce defective blood cells; marrow from a healthy person can sometimes provide a source of healthy ones. A similar strategy has been tried in Parkinson's and Huntington's diseases, which are caused by the death of neurons in specific regions of the brain. Stem cells might grow into healthy neurons that can at least delay the course of the disease. The main problem is obtaining and keeping certain types of adult stem cells. They are rare, and when extracted from the body, they almost immediately develop into specialized types.

In organ transplants, it is often difficult to find a donor who matches a patient; the same is true of cell transplants. Injecting one person's cells into someone else would probably trigger an immune reaction and possibly disease or death. This problem could be overcome by using a person's own cells. However, it is extremely difficult to find and remove adult stem cells from a human being, and even when possible, if the problem is a genetic disease, these cells are likely to be defective, too.

Samuel Wood's success in creating clones from nuclei taken from skin cells suggests that it may be possible to create embryos from adult cells and use them to make new stem cells. Otherwise, a current practice is for hospitals to collect and freeze tissue taken from the umbilical cord of newborns, which is rich in embryonic stem cells. The idea is to keep this tissue until a person might need it, then to use it to produce various types of more specialized stem cells.

In either case, researchers may need to work with cells that might develop into a full human being if they were implanted into a woman's womb. Thus, some people consider experiments with embryonic stem cells to be the moral equivalent of conducting experiments on human beings. This issue was introduced above in the section on reproductive cloning, but it bears reconsidering, because surveys show that people have different attitudes about the two types of cloning. They may be unsure about the meaning of the terms. For instance, a Gallup poll from May 2002 showed that 59 percent of the Americans who were surveyed approved of the "cloning of human organs or body parts that can then be used in medical transplants," but only 34 percent approved of the "cloning of human embryos for use in medical research." And 51 percent approved of the "cloning of human cells from adults for use in medical research." It is unclear whether people are aware that it may be necessary to make "embryos" to obtain such cells.

Many researchers argue that cells that will never be put into a womb to develop will never become human, so they are no different than blood samples or other types of tissue grown in cultures in hospitals and laboratories throughout the world. If the potential to become human were enough to give cells the same rights as fully developed human beings, then there would have to be new rules regarding the handling of single sperm and egg cells, which also—under the right circumstances—might become human beings. Obviously, the question of abortions would need to be handled differently as well. Under the current laws of the United States and many other countries, a pregnancy may be terminated during the first three months of an embryo's development. After that, a fetus has a special status under the law and may not be aborted except in carefully defined circumstances. This is a practical solution to a very complex ethical issue involving the definition of human life, the rights of mothers to make decisions regarding their bodies and their health, and the protection of the unborn, but it has left many people unhappy, including religious leaders.

In the United States, experiments involving human cells or tissues have to be approved by ethics committees. A scientist has to demonstrate that there is a need for the experiment, that it cannot be done using other methods, and that ethical considerations have been taken into account. The most common source of embryonic stem cells are frozen embryos that have been created and maintained by fertility clinics for in vitro fertilization (IVF). In trying to help families that are otherwise unable to have children, such clinics produce far more embryos than can ever be implanted or used. According to statistics from the

Human Fertilisation and Embryology Authority (HFEA) in the United Kingdom, a country in which stem cell experimentation is permitted, a total of 925,747 embryos were created between 1991 and 2000 by clinics for use in IVF. Only 3.5 percent of these were successfully implanted in mothers and survived to birth. All the rest, except for 118 embryos that were donated to research, were simply destroyed. Advocates of stem cell research argue that it would have been better to put the cells to use in research that might one day benefit other people.

It is possible to create embryonic stem cells engineered in such a way that they could never become human beings. For example, scientists could delete an essential protein, such as a molecule needed for the embryo to attach itself to the lining of the mother's womb. The stem cell would still be useful in therapies, but it could never become human; some feel that this eliminates the moral issue.

Most scientists agree that it would be unethical to manipulate the human germline—the reproductive cells that go on to make humans. So the focus of molecular medicine, described in the last chapter, has been to develop therapies that can be used to cure genetic defects in adults or fully developed embryos. Some of the most promising methods involve siRNAs, or gene therapies involving viruses (discussed in the previous chapter). Most people see these methods as similar to normal medicines and do not have ethical problems with their use.

GENETIC TESTING AND CONCERNS ABOUT EUGENICS

Many people's concerns about cloning are tied to other issues, particularly worries that genetic tests or engineering will be used to change or "improve" the human species. In a survey conducted by the Discovery Channel in 2002, 87 percent of Americans disagreed with the statement, "Parents should be allowed to use gene technology to 'design' a baby to satisfy their personal, cultural or aesthetic desires." On the other hand, they did not consider it as bad to use genetic tests on embryos to select desirable characteristics. In the survey, 42 percent agreed that "Parents should have the right to screen out embryos that are found to be carrying a hereditary disease, so that only those free from the condition are allowed to be born," and 48 percent agreed that "Parents should be allowed to select an embryo in order to help cure a sibling of a serious disease." A wide range of new genetic tests are making such choices possible.

From the late 19th century until about 1940, some of the world's most prominent scientists and philosophers—including Francis Galton (a cousin of Charles Darwin), U.S. president Theodore Roosevelt, and David Starr Jordan, the president of Stanford University—suggested that humans should take control of their own evolution. Eugenics movements aimed to use genetic principles and breeding practices to improve the species at a time when very little was actually known about human genetics. The positive form of eugenics encouraged smart and wealthy people to mate with each other, hoping that future generations would produce more smart and wealthy people. Negative eugenics aimed to eliminate "bad" genes by sterilizing or even killing those with disabilities or undesirable social characteristics. The United States and many other countries sterilized thousands of people in mental health institutions, criminals, alcoholics, and others. The most extreme case was the Holocaust, an atrocity the Nazis tried to justify with eugenics arguments. There was almost no scientific basis for any of these attempts to improve the human race. Eugenics was based on a vastly oversimplified way of thinking about genes and very subjective, culturally bound prejudices about what makes some human beings better than others. The abuses were so horrible that, today, societies across the world generally reject the idea of trying to improve humanity by making direct changes in the human genome.

Except in very rare cases involving severe genetic diseases, it is not possible to use information from the genome to make predictions about a baby's future intelligence, behavior, or lifestyle. Yet for some reason, there is a very widespread idea that genes have a deterministic influence and take away human freedom. When a few studies appeared stating that certain genetic characteristics could be linked to violent behavior—at a low statistical rate—lawyers immediately saw an opportunity for a new type of defense. They claimed that their clients were powerless to overcome the influence of their genes, even though there was plenty of scientific evidence to show that the environment played a huge role in whether people with certain genes became violent.

This is an echo of eugenics, and it can be found in nearly every situation where people make assumptions about others based on genetics. When families with fertility problems look for sperm or egg donors, they frequently advertise for artists, scientists, wealthy people, or donors with high scores on college entrance examinations, hoping that some of these qualities will be passed along to the child in the genes. There are no genetic tests that can determine whether one sperm or egg will produce a smarter, better behaved, or more talented baby than another.

Thus, today, the idea of improving humanity through breeding or genetics is regarded as ethically unacceptable almost everywhere. Yet most people would like to see cures for genetic diseases that cause death or a great deal of suffering.

Approximately 10,000 diseases are now known to be directly caused by defects in single genes or a problem with a chromosome. Some examples are hemophilia, cystic fibrosis, and muscular dystrophy. Each of these is rare, but taken together, about 5 percent of people suffer from one of them. Some forms of genes do not always directly cause a disease, but if a person has them, he or she is at greater risk than the rest of the population. For example, a woman who has inherited a defective form of the BRCA1 gene is 85 percent more likely to develop breast cancer than those who do not.

If such a disease appears frequently in a family, there may be a test available to discover whether an adult, child, or embryo is a carrier of the disease. By 2003, about 200 such tests were available. Discovering one of these genes in a patient can help a doctor take preventive measures, recommend changes in a person's lifestyle, or find the right medication to treat someone.

Another example is Tay-Sachs disease (TSD), common in people descended from Ashkenazi Jews (a group that lived in

Germany in the Middle Ages), French Canadians, and the Cajuns of southern Louisiana. Centuries ago, an ancestor in each of these populations suffered a mutation in a gene called HEXA, and it has been passed down to generations ever since. The defect is recessive, and people with one copy of the gene show no obvious symptoms. Genetic tests can reveal whether an embryo has two copies. If so, the child will suffer from a devastating illness because his or her body is unable to process fats. Molecules called *lipids* accumulate in the brain, and the children nearly always die in infancy after a great deal of suffering.

In other cases, it is clear that an unborn child will not develop normally, but it is impossible to predict the quality of life that he or she will have. A good example is Down syndrome, which appears when a person inherits an extra (third) copy of chromosome 21. People with the syndrome usually have a variety of health problems, including learning disabilities. Yet there is no way to foresee how severe the problems will be in an individual case, and the environment plays a huge role in the mental and social development of such children. A caring home environment can often help them become integrated into society and lead happy lives.

The diagnosis of a genetic disease usually has a huge impact on patients and their families. Parents-to-be may discover that their baby will have a serious health problem. Adults may learn that they are likely to suffer from Parkinson's or Alzheimer's disease—either through tests conducted on themselves or other members of their families. *Genetic counseling* educates people about the causes and transmission of genetic diseases and the risks associated with them.

A counselor usually starts by assembling a *pedigree* (a family tree showing patterns of who has been affected by a particular disease and who has not). From this information alone, a doctor may be able to say that someone is not a carrier; in other cases, there is a risk, and further tests are necessary. Most cases are not this straightforward because two or more genes contribute to the development of a disease. Here it is much more difficult to draw a pedigree, analyze the problem, and predict the *pene-*

trance of the disease (whether a person will show mild or severe symptoms, and at what age they will appear). Yet genetic testing can provide a crucial early warning. Some conditions can be treated with medications or a special diet if a problem is detected early enough. In other cases, tests help prepare families psychologically for the birth of a child with health problems.

Genetic tests may be noninvasive (for example, examining a blood sample), or they may require samples of tissue to be taken from an expectant mother or her child. In a small percentage of cases, the tests produce false positives: An embryo appears to have a defect that is not really there. This is more likely to happen with noninvasive procedures. Testing a mother's blood for proteins linked to Down syndrome is accurate more than 90 percent of the time. If the test is positive, doctors usually follow up with amniocentesis, an invasive procedure that requires drawing fluid from the mother's womb; this procedure is accurate more than 99.8 percent of the time for Down syndrome.

If there is a strong indication that a child will suffer or die from an incurable disease, some parents choose to have an abortion, but this is an individual choice that must be made in accordance with their country's laws. Genetic counselors are trained to be nondirective. Once the test results are known and the family has been provided with facts, the counselor remains on hand to answer questions but does not give direct advice.

Opinions and regulations concerning abortion vary widely across the world, and the subject is a thorny ethical issue almost everywhere. Some people completely reject it for religious or moral reasons. Churches have a variety of stances on the issue. The Roman Catholic Church not only objects to abortion but also all forms of birth control (except abstinence). Other religious groups have more moderate stances.It is outlawed in some countries and permitted under certain conditions in countries such as the United States.

In some places, it has even been encouraged as a form of birth control, and many people are concerned that the widening number of genetic tests will lead to more abortions as a means of selecting "better" children. Would this actually happen? In the 1970s, the People's Republic of China began a campaign to reduce the country's exploding population by limiting the size of families. Policies varied from place to place, but in principle, especially in cities, couples were strongly encouraged to have only one child. If a woman became pregnant with her second child, there could be legal and financial penalties (for example, the family might have to pay a fine or take over their own health care payments). In rural areas, families were sometimes permitted to have a second child if the first were female or disabled. In some cases, women were forced to undergo abortions. Recently, the rules have been relaxed in some regions because of negative population growth and concerns that there will not be enough young workers to support a growing number of retired and elderly people.

When the Chinese policy was put into place three decades ago, there was worry that parents would prefer boys for traditional and economic reasons and would abort females. If a family could have only one child, they might be more likely to abort other types of embryos as well. At that time, the tests that could be carried out on a fetus were much more limited than they are today. It was possible to determine the embryo's sex, and to some extent, this led to the selective abortion of females. In 2000, China had 117 males for every 100 females. Factors other than abortion may have played a small role—for example, a study from 2005 showed that Chinese women are more likely to die from infections of hepatitis B. But the abortion of girls has certainly contributed. Follow-up studies have shown that most Chinese couples kept their first child, regardless of its sex. The situation was different in cases where couples were allowed to have more children. If the first child was a girl, families often took measures to ensure that the second would be a boy (usually by having abortions). Families who already had two or more boys often chose a girl. Genetic testing is not to blame for this, because most Chinese couples did not have access to these procedures. But if the tests had been available, they undoubtedly would have been used in some of the same ways.

Until recently, the availability and cost of laboratory tests have limited the number of screens routinely carried out on fetuses. That is changing rapidly. Parents who want the "best" child might use a wide range of newly available tests as a basis for abortions—a new form of eugenics. This is ethically troubling, because it abuses the tests by assuming that they give information that they do not. Francis Collins, a scientist and devout Christian who headed the public Human Genome Project, recently pointed out that "Genetic determinism . . . implies that we are helpless marionettes being controlled by strings made of double helices. That is so far away from what we know scientifically!" Parents who expect that a genetic test will provide them with a perfect child will be disappointed—they are likely to expect too much of their offspring.

Starting in 1993, Francis Collins headed the National Center for Human Genome Research, later the National Human Genome Research Institute, which spearheaded the Human Genome Project. (*National Human Genome Research Institute*)

Even if this type of genetic selection becomes more common, it is unlikely to have a very significant influence on human evolution. Overall, families who abort an embryo after a genetic screen will probably have fewer children than those who do not screen, and the winners in evolution are those who have the most children (because a greater proportion of the next generation will bear their genes). Even if a majority of the population were carrying out selective abortions, which is not the case, there would be a problem only if they were all choosing very similar offspring.

Tests conducted purely for medical reasons may also have an ethical dimension. Testing an embryo or a child may unintentionally reveal that one of the parents suffers from an incurable disease or that a baby's legal father is not the biological one. In some countries, people may be obliged to share the results of tests to their insurance companies, without knowing how they will be interpreted or used.

GM FOODS AND THE RISE OF ENVIRONMENTAL MOVEMENTS

A century ago, breeders and geneticists were regarded as heroes and potential saviors of the human race—today they are more likely to be associated with Dr. Frankenstein. Where did this dramatic shift of perspective come from? Part of the reason lies with the history of science in the 20th century. Discoveries in physics produced unimaginably powerful weapons. Politicians and racists misquoted evolutionary theory and misused early ideas about genetics as they carried out eugenics programs that sterilized "unfit" people and carried out mass murder under the guise of creating pure races. It did not help that genetic science became highly specialized. With the discovery of genes and the structure of DNA, the hobby breeders of the 19th century were suddenly replaced by researchers who needed years of training to make contributions to their field.

Genetically modified foods entered the market in the 1990s against a backdrop of new public concerns about technology and the environment. The 1960s and 1970s exposed the dangers of pollution, a product of technology and industrialization. In the 1980s, a huge ozone hole was discovered over Antarctica—partially caused by industrial chemicals—which allowed dangerous solar radiation to penetrate the atmosphere and increased the risk of skin cancer for people living in the Southern Hemisphere. Atmospheric studies warned of a greenhouse effect that could dramatically change the global climate. Asthma, allergies, and cancer were on the rise. Governments began to pass laws to minimize environmental pollution and to ban cancer-causing substances. One was DDT, a chemical that was initially promoted as a miracle pesticide and

solution to disease. It quickly acquired a completely different reputation.

DDT played a central role in the development of a large environmental movement in the United States. In 1962, American biologist Rachel Carson wrote a popular book called *Silent Spring* citing scientific evidence that DDT could cause cancer and other types of environmental damage. The book and careful scientific studies demonstrated that DDT was absorbed by crops and entered the diets of animals and humans. It had a negative effect on birds, fish, trees, and many other forms of life. The public became alarmed and put pressure on the government; the substance was soon banned in the United States and many other countries. Yet the ban also had negative consequences and became a complicated ethical issue: DDT had saved millions of lives by killing mosquitoes and other insects bearing malaria, typhus, and other diseases. It was the main tool in the World Health Organization's global campaign to wipe out malaria, which had succeeded in most of North America and Europe. The lives saved had to be balanced against millions of deaths that might have been reduced through the use of the insecticide. DDT had been sold to the public in extensive publicity and advertising campaigns; later it was seen as a toxin. Now GM crops were being promoted; were they really safe?

Initially, the public responded favorably to the Flavr Savr tomato, the first genetically modified food to go on the market. But in many places, particularly Europe, that would soon change. *(Emerald Insight)*

Another factor in the public's response to GM crops was a recent outbreak of mad cow disease (technically known as bovine spongiform encephalopathy, or BSE). While they did not know what caused the disease, scientists had good evidence that it could spread when the feed of cattle or other farm animals was contaminated by remains of infected animals (for example, meat and bone meal from other cows). An outbreak in the 1990s caused several deaths in Europe. The British government at first denied that infected animals were to blame. As the situation worsened, the government was blamed for not giving people enough information to protect themselves and for failing to regulate the quality of food. As a result, many people were no longer confident that food, chemicals, or other products of science were adequately tested before going onto the market. GM crops might bring hidden, long-term risks to people's health.

People also worried that modified plants and animals might have an unforseeable impact on the environment. Transplanting new species into the environment had sometimes led to disaster. In the 1960s, scientists introduced a fish called the Nile perch into Lake Victoria in central Africa, the source of the Nile River. The perch was so well suited to its new home and multiplied so quickly that native species of fish have been virtually wiped out. The story is told in a 2005 documentary called *Darwin's Nightmare,* written and directed by the Austrian filmmaker Hubert Sauper. The film explores the human and economic impact of the fish, now such an important source of food that it is bartered to Russian buyers in exchange for weapons. Other cases of transplantation have been unintentional, such as the transport of pests in food containers and the snails or other small animals that are often carried along in the ballast water of ships. When ships travel across the world and illegally empty their tanks, the result may be infestations or the disruption of local food chains.

These situations are not directly related to genetic engineering, but they have set the backdrop for people's responses. While some of the arguments against GM foods or other organisms are based on scare tactics rather than realistic estimations of risk, scientists admit that the effects of GM foods or other organisms are impossible to predict with absolute certainty. Every organism lives in a complex network of interactions with every other, from bacteria in the soil to other plants and animals. Testing a new strain's effects on all of them would be impossible.

Legally, however, the question became whether GM species should have to meet far stricter standards of safety than any other new product brought onto the market. Many people thought so because they felt that genetic engineering was tampering with nature. While farming also altered organisms through selection, changing their genomes in unpredictable ways, there was a difference. A farmer has to work with the changes in plants and animals that arise through mutations and other natural processes.

New strains created by farmers are not subjected to the same tests, even though they may have undergone many more mutations than a GM organism, and their effects may never have been studied in the laboratory. The effects are particularly dramatic in hybrids. Today's supermarkets sell bread made from wheat that originated as a cross between two ancient species of grass; their hybridization produced a huge, double genome, with thousands of extra genes. Natural hybrids may affect the environment as much—or even more—than GM crops; the results are equally impossible to predict in detail. Often, hybrids are much more fertile than the parent strains that produced them, and the chances that they will be disruptive might be very high since so many genes have undergone changes.

In the United States, the Food and Drug Administration is responsible for approving GM foods intended for the market after extensive testing in laboratories. The Flavr Savr tomato, the first such crop, was approved after the FDA determined that it did not pose a health hazard. Producers were not required to give it special labeling. In 2003, a survey conducted by ABC News showed that 92 percent of Americans believed that "the federal government should . . . require labels on food saying whether or not it has been genetically modified or bio-engineered." The percentage had steadily risen since similar surveys in 1998 (82 percent in favor of labeling) and 2000 (86 percent).

Some common fears, such as the idea that modified genes from a plant might enter a person's body and cause health problems, are simple misunderstandings about how genes work. Scientists have never discovered a case where a gene from a plant has been taken up by the human genome by eating; all food contains foreign DNA, and it is destroyed during digestion. Some of the concerns may have arisen because of mad cow disease, but there the cause is a protein fragment; genes do not behave the same way. Another source of confusion might be the publicity surrounding horizontal gene transfer (HGT), in which bacteria or viruses capture genes from one host and transmit them to another. Scientists are unsure how often this happens in nature, but once again, the cause is not digestion.

While most scientists admit that it is impossible to calculate all the risks involved in the creation and spread of GM organisms, they are concerned that the debate has not been balanced and sufficiently informed by facts. Many feel that science fiction movies, negative publicity, and misunderstandings have given the public a false idea of what GM organisms are and how risky their use might be. Just as the media present much more bad news than good, problems with GMOs receive far more attention than the positive effects they have had on millions of people's lives. Responses to applications of genetic engineering have been partially conditioned by people's feelings about other things, such as their opinions about the behavior of governments and businesses. At the same time, GMOs are not simply a matter of science. They are also products, and whether good or bad things happen to them depends on the behavior of the businesses that market them and the people who use them.

The reaction to GM foods has puzzled researchers because many people who are afraid of them do not object to other things that are much more dangerous. There is no evidence that GMOs come with greater risks than the plants and animals produced by traditional farming and breeding, unless an organism has qualities that make it much better at survival and reproduction than other similar strains. GMOs have been designed to create new foods, curb hunger and starvation, and prevent disease. Not using science to try to solve some of these very grave problems would be ethically very questionable. Discussions should not be entirely focused on risks; one must also weigh the potential benefits and give equal consideration to the consequences of not taking action at all.

After all, most of the settled areas of the Earth have been entirely shaped by human "tampering" with the environment. Few of the crops people eat today and very few of the animals resemble the species that inhabited the globe 10,000 years ago, at the dawn of agriculture. The crops of 500 years ago would never have supported the current population of the globe; even today's crops, which have a much higher yield, are unable to support it. As the population continues to rise and resources are exhausted, agricultural science has to continue to adapt. Genetic science arose from an understanding of the complexity of organisms and their interactions with the ecosphere. It may be the only way to feed humanity in the future without causing irreparable damage to the environment.

OWNING GENES, GENOMES, AND LIVING BEINGS

In 1997, Stuart Newman and colleagues at the New York Medical College applied for a patent to obtain ownership of some creatures that were part animal, part human. These organisms do not yet exist, but one day researchers might create something like them. Scientists had already created animals with human genes and humanlike organs. At the University of Nevada, human stem cells had been implanted into sheep embryos with the aim of obtaining livers and other organs that could be transplanted into humans. Pigs had been engineered to produce human hemoglobin, and Stanford University scientists had injected immature human brain cells into mouse fetuses, hoping to create mice that could serve as models of Parkinson's disease and other brain disorders. Embryonic cells from sheep and goats had already been fused to create geeps—half sheep and half goats. These trends worried Newman and his coapplicant, biotechnology activist Jeremy Rifkin. There was nothing to prevent scientists from carrying out similar projects on humans and their closest living relatives, chimpanzees or other primates.

Newman and his colleagues hoped the application would call attention to research that might be legal yet ethically very questionable. A patent would allow them to block other scientists from making such creatures, buying time for scientists and governments to think about the problem. The U.S. Supreme Court had set a dangerous precedent in 1980, they felt, with a decision that allowed the patenting of a microorganism. The U.S. Patent Office had moved further down the slippery slope by granting a patent for a mouse strain to Harvard University. In that case, Philip Leder and his research team had genetically modified the strain by inserting a gene that caused the animals to develop cancer. This onco-mouse has been useful to researchers all across the globe in uncovering some of the causes of cancer, developing drugs to combat it, and detecting carcinogens in the environment. When the patent was awarded, animal rights and environmental groups protested, both on the grounds that the animals suffered and that species should not be ownable. Defenders of patenting said that without the profits that could be gained through ownership, companies would not be motivated to invest in medical research. After a pause of nearly five years, during which experts tried to untangle the legal and ethical issues surrounding

Jeremy Rifkin, biotechnology activist, participated in a patent application to obtain ownership of animals that would be part animal, part human, in order to demonstrate gaps in the regulations covering ownership of genetically modified organisms. *(Vegan Underground)*

the case, the Patent Office once again began issuing patents on genetically modified animals.

The application made by Rifkin and Newman was rejected on the grounds that it proposed creating animals that were "too human." Yet other patents had already been granted concerning animals with human genes or with humanlike organs. Obviously, there were problems: where to draw the line between the human and the nonhuman, and why some forms of life should be ownable but others not. These questions are not well defined in current laws or patent practices.

Another type of ownership question arose in Iceland in 1996 with the formation of deCODE Genetics, Inc. Here the issue was information, not species themselves. The company hoped to find diseases caused by combinations of genes, rather than single mutations—an extremely difficult problem that is best approached by looking at data from huge numbers of patients with extensive medical records and clear family relationships. Iceland seemed to offer a unique opportunity: Since its settlement in the ninth century, the country's population has been isolated to an unusual degree. It is easiest to do genetic studies with very homogenous populations, and in Iceland the descendants of the original inhabitants were thought to have intermarried, with little influx of foreign genes. Finally, detailed genealogical records had been kept for centuries, tracking births and family histories. These factors seemed to make the 288,000 inhabitants of the country ideal subjects in which to hunt for multigene diseases.

The project needed extensive medical records from the population, so deCODE proposed the creation of a national health sector database that would collect information from the medical and genealogical records of all Icelanders. This idea provoked an ethical controversy, however, since it would obviously be impossible to obtain permission directly from everyone, which is normally required by law. The Icelandic legislature passed a bill in 1998 to permit the project to go ahead without citizens' consent, but five years later it was overturned by the country's supreme court on the basis that the database violated their rights. The company plans to continue to work on the same scientific questions, taking a different approach. Its strategy has also changed with the discovery that the Icelandic genome has not been as isolated from external influences as was believed.

The controversy over the ownership of species and biological information is a good example of the challenge that the pace of science poses to the legal system and society. Understanding and curing many diseases may well depend on amassing information that is now considered private and unavailable. On the other hand, no one knows whether the approach of deCODE will lead to cures. So the issue is not simply a question of balancing individual rights against the common good: The court ruling states that those rights must take precedence over a hope for medical progress. It would be interesting to know how the court would have responded if deCODE had been able to guarantee success.

Scientists and doctors are not the only people who seek access to privileged medical information. Businesses would like to know if their employees are likely to become sick, or if they have allergies related to the workplace (which people might not know themselves). Insurance companies are successful only because they can make predictions about the health of their clients. If it were possible to use genetic tests to do so, they might use the information to charge higher rates or even to refuse to offer insurance to someone whom they considered to be a risk.

A LOOK FARTHER AHEAD

The science of evolution demonstrates that merely by living together, species change one another. This book began by showing how humans were the first to take deliberate control of that process by domesticating and breeding plants and animals. For more than 10,000 years, this process was limited because farmers and breeders could work only with varieties that nature provided. Mendel's laws, the discovery of the structure of DNA, and the development of genetic engineering have now given humankind the tools to take control of other species and to intervene directly in their evolution. And now, at the beginning of the 21st century, humans are beginning to have a more direct impact on their own nature. The first step has been to make life more comfortable by changing other creatures—lessening the threat from viruses and other infectious agents with new medicines, improving the nutritional value of plants, and turning animals into drug factories or models that can be used to study human diseases. The next step might be to intervene directly in the human genome.

This process has already begun with the development of new types of therapies that seek to correct defective genes. The current strategies are experimental and limited, but they will quickly improve. Scientists believe that they will learn to deliver foreign genes, RNAs, or proteins to the right targets in the body, and then an entirely new era of medicine will begin. This will not remove the huge amount of chance involved in heredity, but it will mean that people no longer have to suffer all the consequences of "reproductive roulette." They will be able to modify some of the genes that they were born with by intervening in processes going on in their cells.

All of these things may be possible, but will they happen? How far will they go? Surgery began as a last resort to save lives, but today it is widely used as an aesthetic tool by which people try to make themselves look younger or more attractive. Modern drugs were developed to cure disease; now they are used to help people win bicycle races or hit more home runs, to calm excitable children, and even for entertainment. Can there be any doubt that genetic engineering will be used the same way? If it were possible to double lifespans, to regrow lost limbs or damaged brain cells, or to make people healthier, happier, and more intelligent, who would stand in the way? What is possible is not always desirable or ethical, but those judgments change with time.

Today many people find the idea of altering human nature shocking and immoral. Part of this reaction has to do with serious ethical concerns, but another part may be due to the fact that genetic engineering is so new and that it is not yet possible to do things safely and well. And the horrors of negative eugenics—attempts to improve the human race by sterilizing or killing millions of people that politicians found undesirable—lie just a few years in the past. But the Holocaust had almost nothing to do with science; it was brought about by a society plagued with racism and entranced by a cult of power. Science can become a tool of power and brutality; religion and ideas can also be used in this way. Aldous Huxley, whose book *Brave New World* portrays a dark future in which humans have been specially bred to work as slaves, made these points when asked about progress in biology. He said he had nothing against genetic engineering in itself—the new subspecies of humans in *Brave New World* could also be created without it.

How will the genetic engineering of humans be regarded in the future? Will it be viewed as a luxury? Gene therapy clinics may one day remove the need for plastic surgery, tattoo parlors and tanning salons, nursing homes, and psychiatric wards. People may consult genetic counselors the way they use personal fitness trainers—to get advice on the changes they wish to make in their bodies, their lifestyles, and even their personalities.

If these things seem unlikely, consider a poll carried out in 2002 for the Genetics and Public Policy Center, in which Americans were asked whether they approved of the following uses of preimplantation genetic diagnosis (PGD)—in other words, whether embryos produced in a test tube should be screened in the following ways before being implanted in a mother's womb:

- use PGD to make sure their baby does not have a serious genetic disease (74 percent approved)
- use PGD to make sure their baby does not have a tendency to develop a disease like cancer when he or she is an adult (60 percent approved)
- use PGD to make sure their baby would be a good match to donate his or her blood or tissue to a brother or sister who is sick and needs a transplant (69 percent approved)
- use PGD to choose the sex of their child (28 percent approved)

- use PGD to make sure their baby has desirable characteristics such as high intelligence and strength (22 percent approved)
- change their own genes in order to prevent their children from having a genetic disease (59 percent approved)
- change their own genes in order to have children who would be smarter, stronger, or better looking (20 percent approved)

Science fiction author Michael Crichton points out that visions of the future are almost always wrong. People's lifestyles today are dominated by inventions and technology that often arose by accident, completely unexpectedly, in areas that no one even considered a few decades ago. When it comes to genetic engineering, futurologists have predicted that humans will be engineered to grow tentacles or extra eyes, to survive underwater or in outer space, to connect their minds directly to computers or their home entertainment centers, or to live for hundreds of years. Any of these things could happen—providing someone sees the point in having tentacles—if democratic societies allow them to. On the other hand, genetic science may run up against an unanticipated barrier. It has not yet been possible to produce cold fusion, to build a spaceship that comes anywhere near the speed of light, or to travel back in time. Single cells—let alone whole organisms—are so complex that it might not be possible to build a computer that can simulate one well enough to predict what will happen if many different genes are changed. The problem may be as difficult as trying to predict the weather months or years in advance. Then again, weather cannot be genetically engineered or simulated in animal models to study how it behaves under controlled conditions.

Where will genetic engineering lead? Most researchers hesitate to make predictions. Two decades ago, scientists predicted that genetics and molecular biology would soon cure cancer, dramatically extend people's lives, and solve a wide range of other problems. Those miracles have not (yet), come to pass, but there have been others. DNA sequencers have finished hundreds of genomes, discovered new diseases, invented a wide range of tools to manipulate genes, and uncovered ancient genetic programs that have steered evolution for more than a billion years. Bacteria and other organisms serve as factories to produce human molecules that are used in therapies. When a new virus arises, its DNA and RNA can be decoded almost instantly, and the information can be scanned for weak points. Epidemiologists can trace the route that HIV has taken from person to person. Forensic scientists can use genetic information to connect suspects to crimes or demonstrate their innocence. Progress over the past few decades has been astounding, and there is every reason to think that it will continue into the future at a faster and faster pace.

Humans and their characteristics, including activities such as science, are a product of evolution. This has implications for how we ought to think about genetic engineering, according to biologist Hubert Markl, former president of the German Research Council and of the Max Planck Society. In an article published in 2002 in the *Journal of Molecular Biology,* he wrote the following:

> We should see ourselves not as some kind of fallen angel, alien intruder, some aberrant or deranged scourge of nature, but as its constituent and heir. And not only as one constituent part of nature among many others, just an arbitrarily chosen biological species, but as a unique, a quite extraordinary kind of natural species, through which nature entered into an entirely new stage of evolution; a species that not only participates in its future evolution like any other species, but that increasingly commands and determines this future, for better or worse. In evolving the human species, nature, as it were, began to take control of its own future, to give it purposeful direction, to assume responsibility for its own future development. . . . From such a comprehensive evolutionary perspective, human technological and economic inventiveness is nothing other than nature's way of intentionally acting upon itself and forming its own future.

It is impossible to predict how people in the future will feel about genetic engineering, or what place it will have in their society. People will have their own values and will no longer care about ours, any more than we base our decisions on the wishes and desires of the kings of the 18th century. Values in the future will be deeply connected to the type of world people live in, and they will use all the tools at hand to survive. Their world may be much like ours, and they may think about it in ways that we would understand. That will probably not be the case if they inherit a planet that is tremendously overcrowded, unable to feed itself, overpolluted, overheated through global warming, irradiated because of large gaps in the ozone layer, or torn apart by wars brought about by huge differences in standards of living across the globe. In any of these worst-case scenarios, the survival of the human race might depend on making some drastic changes in the human genome. That may or may not be desirable. Hopefully, it will never be necessary.

Chronology

1651	William Harvey claims that all animals arise from eggs.
1677	Antoni van Leeuwenhoek discovers sperm.
1694	Rudolph Camerarius identifies the sex organs of plants.
1740	Carl Linnaeus begins to pollinate plants artificially.
1751	Pierre-Louis Maupertuis studies polydactyly, the inheritance of extra fingers in humans.
	Joseph Adams recognizes the negative hereditary effects of inbreeding.
1824	Joseph Lister builds a new type of microscope that removes distortion and greatly increases resolution.
1827	Karl Ernst von Baer is first to discover an egg cell in a mammal (a dog).
1830	Giovanni Amici discovers egg cells in plants.
1838	Matthias Schleiden states that plants are made of cells.

1840	Theodor Schwann states that all animal tissues are made of cells.
1855	Rudolf Virchow states the cell doctrine: All cells arise from preexisting cells.
1856	Gregor Mendel begins experiments on heredity in pea plants.
1857	Joseph von Gerlach discovers a new way of staining cells that reveals their internal structures.
1858	The theory of evolution is made public at a meeting of the Linnean Society in London with the reading of papers by Charles Darwin and Alfred Russel Wallace.
1859	Charles Darwin publishes *On the Origin of Species.*
1865	Gregor Mendel presents his paper "Experiments in Plant Hybridization" in meetings of the Society for the Study of Natural Sciences in Brno, Moravia. The paper outlines the basic principles of the modern science of genetics.
1866	Mendel's paper is published in the Proceedings of the Brno Society for the Study of Natural Sciences but receives little recognition.
1868	Friedrich Miescher isolates DNA from the nuclei of cells; he calls it nuclein.

1871	Francis Galton carries out experiments in rabbits that disprove Darwin's hypothesis of how heredity functions.
1876	Oscar Hertwig observes the fusion of sperm and egg nuclei during fertilization.
1879	Walther Flemming observes the behavior of chromosomes during cell division.
1885	August Weismann states that organisms separate reproductive cells from the rest of their bodies, which helps explain why Lamarck's concept of evolution and inheritance is wrong.
1888	Weismann tries and fails to observe Lamarckian inheritance in the laboratory by cutting off the tails of mice for many generations.
1900	Hugo de Vries, Carl Correns, and Erich von Tschermak-Seysenegg independently publish papers that confirm Mendel's principles of heredity in a wide range of plants.
	Archibald Garrod identifies the first disease that is inherited according to Mendelian laws, which means that it is caused by a defective gene.
	Theodor Boveri demonstrates that different chromosomes are responsible for different hereditary characteristics.
1902	William Bateson popularizes Mendel's work in a book called *Mendel's Principles of Heredity: A Defense.*

1903	Walter Sutton connects chromosome pairs to hereditary behavior, demonstrating that genes are located on chromosomes.
1905	Nettie Stevens and Edmund Wilson independently discover the role of the X and Y chromosomes in determining the sex of animals.
1906	William Bateson discovers that some characteristics of plants depend on the activity of two genes.
1908	Archibald Garrod shows that humans with an inherited disease are lacking an enzyme (a protein), demonstrating that there is a connection between genes and proteins.
1910	Thomas Hunt Morgan discovers the first genes in fruit flies when a screen reveals a mutation that is inherited in Mendelian ratios.
1911	Morgan discovers some traits that are passed along in a sex-dependent manner and proposes that this happens because the genes are located on sex chromosomes. He proposes the general hypothesis that traits that are likely to be inherited together are located on the same chromosome.
1913	Alfred Sturtevant constructs the first genetic linkage map, allowing researchers to pinpoint the physical locations of genes on chromosomes.

1922	Ronald A. Fisher uses mathematics to show that Mendelian inheritance and evolution are compatible.
1927	Herman J. Muller shows that X-rays can cause mutations.
1928	Fredrick Griffith discovers that genetic information can be transferred from one bacterium to another.
1931	Barbara McClintock shows that as chromosome pairs line up beside each other during the copying of DNA, fragments can break off one chromosome and be inserted into the other in a process called recombination.
1933	Theophilus Painter discovers that staining giant salivary chromosomes in fruit flies reveals regular striped bands.
1934	Calvin Bridges shows that chromosome bands can be used to pinpoint the exact locations of genes.
1935	Calvin Bridges and Hermann Muller discover independently that a fly mutation called Bar is caused by the duplication of a gene.
1937	George Beadle and Boris Ephrussi show that genes work together in a specific order to produce some features of fruit flies.
1941	George Beadle and Edward Tatum propose that each gene is responsible for the activity of one enzyme.

1943	Max Delbruck and Salvador Luria demonstrate evolution in the laboratory by showing that bacteria evolve defenses to viruses through mutations that are acted on by natural selection.
1944	Oswald Avery, Colin MacLeod, and Maclyn McCarty show that genes are made of DNA. Erwin Schrödinger publishes *What Is Life?*
1946	Joshua Lederberg and Edward Tatum discover conjugation in bacteria.
1950	Barbara McClintock publishes evidence that genes can move to different positions as chromosomes are copied.

Erwin Chargaff discovers that in DNA samples from different organisms, the base adenine always occurs in the same amounts as thymine and that the same is true for guanine and cystine. |
| **1951** | Rosalind Franklin uses X-ray diffraction to obtain images of DNA; the patterns reveal important clues to the building plan of the molecule. |
| **1953** | James Watson and Francis Crick publish the double-helix model of DNA, which explains both how the molecule can be copied and how mutations might arise.

Rosalind Franklin and Maurice Wilkins publish X-ray studies that support the Watson-Crick model. |

1958   Francis Crick describes the "central dogma" of molecular biology: DNA creates RNA creates proteins. He challenges the scientific community to figure out the molecules and mechanisms by which this happens.

1959   Marshall Nirenberg, Marianne Grunberg-Manago, and Severo Ochoa show that the cell reads DNA in three-letter "words" to translate the alphabet of DNA into the 20-letter alphabet of proteins.

1960   Richard MacNeish discovers traces of ancient maize cultivation in the Valley of Tehuacán, Mexico.

1961   Sidney Brenner, François Jacob, and Matthew Meselson discover the molecule messenger RNA.

Françis Jacob and Jacques Monod demonstrate that the activity of genes is controlled by nearby DNA sequences called operons.

1966   Marshall Nirenberg and H. Gobind Khorana work out the complete genetic code—the DNA recipe for every amino acid.

1970   Hamilton Smith and Werner Arber discover the first restriction enzyme, a molecule in bacteria that cuts DNA.

1972   Janet Mertz and Ron Davis use restriction enzymes and DNA-mending molecules called ligases to carry out the first recombination: the creation of an artificial DNA molecule.

1973   Annie Chang and Stanley Cohen use genetic engineering for the first time to combine DNA from a bacterium and virus and transfer genes between species.

1977   Walter Gilbert and Frederick Sanger devise new methods to analyze the sequence of DNA, launching the age of high-throughput DNA sequencing.

Phillip Sharp and colleagues discover introns, information in the middle of genes that does not contain codes for proteins and must be removed before an RNA can be used to create a protein.

Frederick Sanger finishes the first genome, the complete nucleotide sequence of a bacteriophage.

The first biotech company, Genentech, is founded, which plans to use genetic engineering to make drugs.

1978   Recombinant DNA technology is used to create the first human hormone.

1981   Three laboratories independently discover oncogenes, proteins that lead to cancer if they undergo mutations.

1985   Kary B. Mullis publishes a paper describing the polymerase chain reaction, an easy method for cloning DNA molecules.

1986   First outbreak of BSE (mad cow disease) occurs among cattle in the United Kingdom.

1988	The Human Genome Project is launched by the U.S. Department of Energy and the NIH with the aim of determining the complete sequence of human DNA.
1989	Alec Jeffreys discovers regions of DNA that undergo high numbers of mutations. He develops a method of DNA fingerprinting that can match DNA samples to the person they came from and can also be used in establishing paternity and other types of family relationships.
1990	W. French Anderson carries out the first human gene replacement therapy to treat an immune system disease in four-year-old Ashanti DeSilva.
1993	The company Monsanto develops and begins to market a genetically engineered strain of tomatoes called Flavr Savr.
1995	The first confirmed death from Creutzfeldt-Jakob disease, the human form of BSE, is reported in the United Kingdom.
1996	Researchers complete the first genome of a eukaryote, baker's yeast. The completion of the genome of *Methanococcus jannaschii,* an archaeal cell, confirms that archaea are a third branch of life, separate from bacteria and eukaryotes.
	Gene therapy trials to use the adenovirus as a vector for healthy genes are approved in the United States.

1997	Ian Wilmut's laboratory at the Roslin Institute produces Dolly the sheep, the first cloned mammal.
1998	The first genome of an animal, the worm *Caenorhabditis elegans,* is completed.
1999	Jesse Gelsinger dies in a gene therapy trial, bringing a temporary halt to all viral gene therapy trials in the United States.
2000	The genome of the fruit fly, *Drosophila melanogaster,* is completed.
2001	A complete working draft of the human genome is published.
2002	The mouse genome is completed.
2004	Scientists in Seoul announce the first successful cloning of a human being, a claim that is quickly proven to be false.
2008	Samuel Wood of the California company Stemagen successfully uses his own skin cells to produce clones, which survive five days.

Glossary

allele one variant of a single gene. Humans usually have two copies of each gene, one inherited from each parent, located at the same positions in their two chromosomes, which may be identical or different alleles.

alternative splicing a process by which segments can be removed from immature RNA molecules in different ways, producing different messenger RNAs and thus different proteins

amino acid one of the chemical subunits that make up proteins

apoptosis a built-in genetic program that triggers the death of cells

bacterial artificial chromosome (BAC) an artificial sequence of DNA containing a gene and other elements placed in bacteria for copying

bacteriophage a type of virus that infects bacteria

base pair a unit made of two DNA nucleotides, either an adenosine bound to a thymine, or a guanine bound to a cystine

biometrics the study of any measurable physical characteristics of organisms, such as weight, height, or body fat

bovine spongiform encephalopathy (BSE) a deadly disease in cattle that causes the brain to deteriorate and take on a spongelike texture. Humans can become infected if they eat infected meat; the disease is known as Creutzfeldt-Jakob disease.

chromatin the mixture of DNA and many types of proteins that are commonly attached to it in the cell nucleus

chromosome large, compressed clusters of DNA and many other molecules found in the cell nucleus

cloning a process of creating identical copies of DNA, cells, or organisms, important in biological research (See also REPRODUCTIVE CLONING.)

conditional knockout a method of deactivating genes in an organism only in particular tissues and/or at specific times

denaturing the process of separating two strands of DNA in order to copy or modify it, an important step in DNA sequencing

ddNTP a special version of a nucleotide base, used in DNA sequencing to stop the reaction that copies DNA molecules

discontinuity a concept developed by William Bateson to explain mutations that cause very dramatic, sudden changes in an organism. He regarded this as the probable cause of the development of new species, rather than the gradualist views of Darwin.

DNA (deoxyribonucleic acid) the molecule that contains the hereditary information of all species. DNA encodes RNA molecules, many of which are used to produce proteins.

DNA fingerprinting a method that compares quickly changing regions of DNA from different samples to determine whether they came from the same organism or to determine how closely related two organisms are

endosymbiant an organism that lives inside another, creating a mutually dependent relationship

enzyme a protein that carries out certain types of chemical reactions, such as cutting other molecules or activating them

exon a DNA sequence that contains the information necessary to make a protein sequence

exon junction complex (EJC) a group of proteins that a cell deposits on sites where an intron has been removed from an RNA molecule

founder   the first member of a species to experience a specific mutation that is subsequently passed to its offspring through heredity

gemmules   a hypothetical "particle of inheritance," usually thought to exist in body fluids and that was believed to transfer hereditary information between parents and their offspring (before the discovery of genes)

gene   a sequence of nucleotides in a DNA molecule that holds the information needed by a cell to create a protein

gene therapy   methods that attempt to cure genetic or other diseases by introducing healthy versions of genes into the cells of patients

genetically modified organism (GMO)   a virus, cell, plant, or animal that has been changed through genetic engineering

genetic counseling   a process in which a physician studies genetic information about a person in order to explain a potential genetic condition and to estimate the probability that it will develop or be transmitted

genotype   the genetic makeup of an individual; the set of alleles an organism has for particular genes

germ cells   the reproductive cells that become sperm or egg cells as an organism develops

gradualism (gradualist)   the theory that the natural world has been produced through a series of small steps over long periods of time by the same natural laws and forces that act on the world today

gynandromorph   a single organism that contains genes and characteristics of both sexes

hemoglobin   a protein in blood cells responsible for the transport of oxygen

hemophilia   an inherited blood disease in humans that prevents blood from clotting properly. Since the gene is located on the X chromosome, it is passed to children from their mothers.

heterozygous   an organism that has two different alleles for the same gene

homozygous   an organism with two copies of the same allele for a particular gene

imaginal disk   tissue in the early embryo of an insect that will later develop into legs, wings, or other structures in the adult body

insulin   a hormone in animals that is produced in the pancreas and transmits signals to cells about the body's uptake of food. People with diabetes are unable to produce the molecule, or produce it in too low amounts, so their bodies are unable to control the amount of glucose in the blood.

intron   a region of a gene or RNA that does not encode information to make proteins, which is removed by splicing

inversions   regions of DNA that become reversed through errors in copying or cell division

jumping gene   regions of DNA that can move to different positions in chromosomes within a cell (SEE TRANSPOSON)

knock in   a genetic engineering technique that adds an extra gene to a genome

knock out   a genetic engineering technique that removes a gene or replaces it with a nonfunctional form

ligase   an enzyme that can join fragments of DNA together, used to insert genes in genetic engineering

lipid   small fats such as cholesterol or steroids that are the main components of cell membranes

materialism   the philosophy that living processes can be explained in terms of their physical and chemical properties. In evolution, this implies that life could emerge through the normal action of physical and chemical laws on inorganic substances without the help of an additional external force.

meiosis   a type of cell division that produces two cells with one member of each pair of chromosomes. Meiosis is the process by which organisms create reproductive cells with half sets of the genetic material.

microRNA   a small RNA molecule that does not encode a protein but is used by a cell to regulate the use of other RNAs. By binding to another RNA, it calls up cellular machinery that destroys the other molecule so that it cannot be used to make proteins.

minisatellite   a region of DNA, often inside a gene, that changes more rapidly than other parts of the genome

mutation   a change that happens in DNA—usually a gene—when it is not perfectly copied

noncoding sequences   regions of DNA that do not contain instructions for building proteins

nondisjunction   a situation in which chromosome pairs do not separate properly during cell division

nonsense-mediated mRNA decay (NMD)   a process by which cells recognize that RNAs have been improperly built and dismantle them, acting both as a quality control device and a means by which cells regulate the quantity of proteins that are made from particular genes

nucleotide   the fundamental chemical subunit of DNA and RNA molecules

oncogene   a gene in an animal that increases the likelihood of cancer if it undergoes certain types of mutations

pangenesis   Darwin's flawed hypothesis about heredity, in which cells from various parts of the body produce particles called gemmules that collect in the sex organs and mix to create a new plant or animal

pedigree a family tree showing the distribution of particular hereditary traits

penetrance the degree to which a genetic trait influences a cell or organism

phenotype the physical characteristics or behavior of a cell or organism. A phenotype develops flexibly through interactions between genes and the environment. The same genome can produce radically different phenotypes; for example, the complete set of human genes can build neurons or red blood cells.

plasmid a circular strand of DNA in a bacteria, separate from the bacteria's chromosome and copied independently of it. Artificial plasmids are an important tool in genetic engineering.

polymerase an enzyme that strings nucleotides together into chains to make DNA or RNA molecules

promoter a region of DNA where molecules bind to transcribe a gene into an RNA molecule

protein a molecule made up of subunits called amino acids, synthesized by cells using information in genes. Proteins are often called the worker molecules of the cell because of the many different functions they perform.

recombination a process in which the order of DNA sequences change, usually because of breaks that occur in chromosome pairs as the cell divides. Material from one chromosome can then be transferred to the other.

reproductive cloning a method of creating a new organism using only the complete existing genetic material of another organism, for example by transferring the nucleus of an adult cell to an egg and stimulating it to develop into an embryo

ribosome a molecular machine made of RNA and proteins. Its function is to transcribe the information contained in messenger RNA molecules into proteins.

RNA (ribonucleic acid)   a molecule made of nucleotides that is formed through the transcription of information contained in DNA. There are several types, including messenger RNA molecules, which are used as patterns to build proteins.

RNA interference   a process in which two single strands of RNA bind to each other and make a double strand. Cells usually destroy such molecules before they are used to make proteins.

sequence   a sequential list of the nucleotide bases that make up a region of DNA

sickle-cell anemia   a disease caused by mutations in a gene called globin that results in poorly formed red blood cells. While dangerous to human health, the condition offers protection from malaria.

small interfering RNA (siRNA)   an artificial RNA molecule made to function like a microRNA. When it is introduced into a cell, it binds to an RNA with a complementary strand. This causes the RNA to be broken down or blocks translation; either method prevents the RNA from being used to make proteins.

splicing   the process by which introns are removed from the first version of an RNA molecule transcribed from a gene, and exons are joined together to generate a finished messenger RNA. This mRNA serves as the template for synthesis of a specific protein.

stop codon   a three-letter code in an RNA molecule that signals the end of the protein-encoding region; it tells molecules that translate RNAs into proteins where to stop

teosinte   a wild plant that was transformed into modern-day maize through centuries of breeding by ancient American cultures

transgenic   an animal resulting from a gene inserted into the cell from which the animal developed

transposon   regions of DNA that can move to different positions or be inserted at new places in chromosomes within a cell

vitalism a philosophy that holds that life cannot arise or be explained without a special vital force that animates nonliving matter

X chromosome a sex chromosome; in humans, fruit flies, and certain other organisms, females have two X chromosomes, and males have only one

Y chromosome the male sex chromosome in humans, fruit flies, and certain other organisms

Further Resources

Books

Brown, Andrew. *In the Beginning Was the Worm.* London: Pocket Books, 2004. The story of an unlikely model organism in biology, the worm *C. elegans,* and the scientists who have used it to understand some of the most fascinating issues in modern biology.

Browne, Janet. *Charles Darwin: The Power of Place.* New York: Knopf, 2002. The second volume of the definitive biography of Charles Darwin.

———. *Charles Darwin: Voyaging.* Princeton, N.J.: Princteon UP, 1995. The first volume of the definitive biography of Charles Darwin.

Caporale, Lynn Helena. *Darwin in the Genome: Molecular Strategies in Biological Evolution.* New York: McGraw-Hill, 2003. A new look at variation and natural selection based on discoveries from the genomes of humans and other species, written by a noted biochemist.

Carlson, Elof Axel. *Mendel's Legacy: The Origin of Classical Genetics.* Cold Spring Harbor, N.Y.: Cold Spring Harbor Laboratory Press, 2004. An excellent, easy-to-read history of genetics from Mendel's work to the 1950s. Carlson explains the relationship between cell biology and genetics especially well.

———. *The Unfit: A History of a Bad Idea.* Cold Spring Harbor, N.Y.: Cold Spring Harbor Laboratory Press, 2001. An in-depth account of eugenics movements across the world.

Cavalli-Sforza, L. Luca, Paolo Menozzi, and Alberto Piazza. *The History and Geography of Human Genes.* Princeton, N.J.: Princeton University Press, 1994. This lengthy book studies the distribution of particular forms of genes from humans around the globe and shows what they reveal about how modern *Homo sapiens* settled the planet.

Chambers, Donald A. *DNA: The Double Helix: Perspective and Prospective at Forty Years.* New York: New York Academy of Sciences, 1995. A collection of historical papers from major figures involved in the discovery of DNA with reminiscences from some of the authors.

Chargaff, Erwin. "Preface to a Grammar of Biology." *Science* 172, 3984 (1971). This classic article demonstrates that the distribution of nucleotides in the DNA of different species is not random and provides key data to the solution of the structure of DNA.

Darwin, Charles. *The Descent of Man.* Amherst, N.Y.: Prometheus, 1998. In this book, originally published 12 years after *The Origin of Species,* Darwin outlines his ideas on the place of human beings in evolutionary theory. All of Darwin's works are accessible to high-school level readers.

———. *The Origin of Species.* Edison, N.J.: Castle Books, 2004. Darwin's masterpiece, outlining the full theory of evolution for the first time and gathering a massive number of facts in support of it.

———. *The Voyage of the Beagle.* London: Penguin Books, 1989. Darwin's account of his five-year voyage around the globe, during which he collected specimens and made observations that would lead him to the theory of evolution.

Diamond, Jared. *Guns, Germs, and Steel: The Fates of Human Societies.* New York: W.W. Norton & Co., 1998. This is an extremely interesting study of human societies in terms of the interplay of environment, evolution, and culture.

Fruton, Joseph. *Proteins, Enzymes, Genes: The Interplay of Chemistry and Biology.* New Haven, Conn.: Yale University Press, 1999. An in-depth history of biochemistry. This book is quite technical, probably best suited for advanced college students, but it contains fascinating anecdotes about key figures in the history of biology.

Gilbert, Scott. *Developmental Biology.* Sunderland, Mass.: Sinauer Associates, 1997. An excellent college-level text on all aspects of developmental biology, including "evo-devo" issues.

Goldsmith, Timothy H., and William F. Zimmermann. *Biology, Evolution, and Human Nature.* New York: Wiley, 2001. Life from the level of genes to human biology and behavior, from the point of view of evolutionary theory.

Gregory, T. Ryan, ed. *The Evolution of the Genome.* Boston: Elsevier Academic Press, 2005. A collection of technical articles by experts in genome analysis covering topics such as the origins of genomes, genome duplication, etc.

Hall, Michael N., and Patrick Linder, eds. *The Early Days of Yeast Genetics.* Cold Spring Harbor, N.Y.: Cold Spring Harbor Laboratory Press, 1993. A collection of important papers from pioneers in the field of molecular genetics.

Henig, Robin Marantz. *A Monk and Two Peas.* London: Weidenfeld & Nicolson, 2000. A popular, easy-to-read account of Gregor Mendel's work and its impact on later science.

Jones, Stephen, Robert D. Martin, and David R. Pilbeam, eds. *The Cambridge Encyclopedia of Human Evolution.* Cambridge: Cambridge University Press, 1994. A collection of fascinating articles on all aspects of evolutionary theory up to the genome era.

Judson, Horace Freeland. *The Eighth Day of Creation: Makers of the Revolution in Biology.* New York: Simon & Schuster, 1979. A comprehensive history of the science and people behind the creation of molecular biology, from the early 20th century to the 1970s.

Kohler, Robert E. *Lords of the Fly: Drosophila Genetics and the Experimental Life.* Chicago: University of Chicago Press, 1994. The story of Thomas Hunt Morgan and his disciples, whose discoveries on fruit fly genes dominated genetics in the first half of the 20th century.

Kohn, Marek. *A Reason for Everything: Natural Selection and the English Imagination.* London: Faber & Faber, 2004. This book focuses on the lives and work of researchers such as Haldane, Fisher, and Wright, who proved that genetic science was compatible with evolutionary theory during the first half of the 20th century.

Lutz, Peter L. *The Rise of Experimental Biology: An Illustrated History.* Totowa, N.J.: Human Press, 2002. A history of the development of modern biology, putting evolution into the context of what was going on in other areas of the field.

Magner, Lois N. *A History of the Life Sciences.* New York: M. Dekker, 1979.

McElheny, Victor K. *Watson and DNA: Making a Scientific Revolution.* Cambridge, Mass.: Perseus, 2003. A retrospective on the work and life of this extraordinary scientific personality.

Purves, William K., David Sadava, Gordon H. Orians, and Craig Heller. *Life: The Science of Biology.* Kenndallville, Ind.: Sinauer Associates and W. H. Freeman, 2003. A comprehensive overview of themes from the life sciences.

Reeve, Eric, ed. *Encyclopedia of Genetics.* London: Fitzroy Dearborn Publishers, 2001. A collection of articles by a wide range of experts on the major themes of modern genetic science.

Ridley, Mark. *Evolution.* Oxford: Oxford University Press, 2003. An in-depth look at the modern science of evolution, probably best suited for entry-level college students.

Slotten, Ross A. *The Heretic in Darwin's Court: The Life of Alfred Russel Wallace.* New York: Columbia University Press, 2006. An excellent biography providing a particularly interesting look at the unusual course that Wallace's life took after the codiscovery of evolution.

Stent, Gunther. *Molecular Genetics: An Introductory Narrative.* San Francisco: W. H. Freeman, 1971. A textbook for college students on genetic science up to the 1970s by one of the pioneers of molecular genetics.

Tanford, Charles, and Jacqueline Reynolds. *Nature's Robots: A History of Proteins.* New York: Oxford University Press, 2001. The story of the birth of biochemistry, giving a look at the lives and work of the researchers who uncovered the functions of proteins and other biological molecules.

Tudge, Colin. *In Mendel's Footnotes.* London: Vintage, 2002. An excellent review of ideas and discoveries in genetics from Mendel's day to the 21st century.

———. *The Variety of Life: A Survey and a Celebration of All the Creatures That Have Ever Lived.* New York: Oxford University Press, 2000. A beautifully illustrated tree of life classifying and describing the spectrum of life on Earth.

Wallace, Alfred Russel. *Natural Selection and Tropical Nature: Essays on Descriptive and Theoretical Biology.* Boston: Adamant Media, 2005. A reprint of a collection of Wallace's major papers and writings originally published in 1891.

Watson, James D. *The Double Helix.* New York: Atheneum, 1968. Watson's personal account of the discovery of the structure of DNA.

Wolpoff, Milford H., John Hawks, Brigitte Senut, Martin Pickford, and James Ahern. "An Ape or *the* Ape: Is the Toumaï Cranium TM 266 a Hominid?" *PaleoAnthropology* 2006: 36–50. A fascinating look at the way paleoanthropologists analyze teeth and bone structures to draw conclusions about how hominid fossils should be placed in the evolutionary tree.

Web Sites

There are tens of thousands of Web sites devoted to the topics of genetics, genetically modified organisms, evolution, and the other themes of this book. The small selection below provides original articles, teaching materials, multimedia resources, and Web sites providing links to hundreds of other excellent sites. All sites listed were accessed on June 1, 2008, unless otherwise noted.

The American Society of Naturalists. "Evolution, Science, and Society: Evolutionary Biology and the National Research Agenda." Available online. URL: http://www.rci.rutgers.edu/~ecolevol/fulldoc.pdf. A document from the American

Society of Naturalists and several other organizations summarizing evolutionary theory and showing how it has contributed to other fields, including health, agriculture, and the environmental sciences.

The Center for Genetics and Society. "Public Opinion Surveys on New Human Genetic and Reproductive Technologies." Available online. URL: http://www.geneticsandsociety.org/article.php?id=404. This article presents the results of numerous surveys conducted in the United States and elsewhere on topics related to human cloning and stem cell research.

Cold Spring Harbor Laboratory. "Image Archive on the American Eugenics Movement." Available online. URL: http://www.eugenicsarchive.org/eugenics. An archive of images and material concerning eugenics from the Dolan DNA Learning Center of Cold Spring Harbor Laboratory, New York. The DNALC home page at www.dnalc.org has many other excellent resources concerning biology, genomics, and health.

The Genetizen. Available online. URL: http://www.geneforum.org./blog. An excellent blog on topics in genetics, particularly social and legal issues, with links to many other Internet resources.

Maddison, D. R., and K. S. Schulz, eds. "The Tree of Life Web Project." Available online. URL: http://tolweb.org. A site that has collected a huge number of articles and links from noted biologists on the question of assembling a "family tree" of life on Earth.

The National Center for Biotechnology Information. "Bookshelf." Available online. URL: http://www.ncbi.nlm.nih.gov/sites/entrz?db=books. Accessed August 7, 2008. A collection of excellent online books ranging from biochemistry and molecular biology to health topics. Most of the works are quite technical, but many include very accessible introductions to the topics. Some highlights include *Molecular Biology of the Cell, Molecular Cell Biology,* and the *Wormbook.* There are also annual reports on health in the United States from the Centers for Disease Control and Prevention.

National Center for Science Education. "Evolution/Creationism in the News." Available online. URL: http://www. natcenscied.org. The homepage of the National Center for Science Education collects stories in the news related to evolution and creationism.

The National Health Museum. "Access Excellence: Genetics Links." Available online. URL: http://www.accessexcellence. org/RC/genetics.php. Links and resources from the "Access Excellence" project of the National Health Museum.

Social Science Research Council. "Race and Genomics." Available online. URL: http://raceandgenomics.ssrc.org. A Web forum in which experts discuss whether "human races" can be defined biologically in light of information from the human genome.

TalkOrigins. "The Talk Origins Archive." Available online. URL: http://www.talkdesign.org. A Web site devoted to "assessing the claims of the Intelligent Design movement from the perspective of mainstream science; addressing the wider political, cultural, philosophical, moral, religious, and educational issues that have inspired the ID movement; and providing an archive of materials that critically examine the scientific claims of the ID movement."

The Tech Museum of Innovation, San Jose, Calif. "Understanding Genetics: Human Health and the Genome." Available online. URL: http://www.thetech.org/genetics. An excellent collection of news and feature stories on scientific discoveries and ethical issues surrounding genetics.

University of Cambridge. "The Complete Works of Charles Darwin Online." Available online. URL: http://darwin-online. org.uk. An online version of Darwin's complete publications, 20,000 private papers, and hundreds of supplementary works.

University of Utah, Genetic Science Learning Center. "Learn Genetics." Available online. URL: http://learn.genetics.utah.

edu/. Delivers educational material on genetics, bioscience, and health topics. The material meets selected U.S. education standards for science and health.

The Vega Science Trust. "Scientists at Vega." Available online. URL: http://www.vega.org.uk/video/internal/15. Accessed August 7, 2008. Filmed interviews with some of the great figures in 20th-century and current science, including leading figures in evolutionary research.

Index